The Purposeful Universe

The Purposeful Universe

How Quantum Theory
and Mayan Cosmology Explain the
Origin and Evolution of Life

Carl Johan Calleman, Ph.D.

Bear & Company
Rochester, Vermont • Toronto, Canada

Bear & Company
One Park Street
Rochester, Vermont 05767
www.BearandCompanyBooks.com

Bear & Company is a division of Inner Traditions International

Library of Congress Cataloging-in-Publication Data
Calleman, Carl Johan.
  The purposeful universe : how quantum theory and Mayan cosmology explain
the origin and evolution of life / Carl Johan Calleman, Ph.D.
      p. cm.
  Includes bibliographical references and index.
  Summary: "Identifying the Mayan World Tree with the central axis of the
cosmos, the author shows how evolution is not random"—Provided by publisher.
  ISBN 978-1-59143-104-6 (pbk.)
  1. Evolution (Biology)—Philosophy. 2. Quantum theory. 3. Maya cosmology. I.
Title.
  QH360.5.C354 2009
  576.801—dc22

                                                                    2009034360

Printed and bound in the United States by Lake Book Manufacturing

10 9 8 7 6 5 4 3 2 1

Text design and layout by Jon Desautels
This book was typeset in Garamond Premier Pro with Modula Tall as the display
typeface

To send correspondence to the author of this book, mail a first-class letter to the
author c/o Inner Traditions • Bear & Company, One Park Street, Rochester, VT
05767, and we will forward the communication.

To my mother

Contents

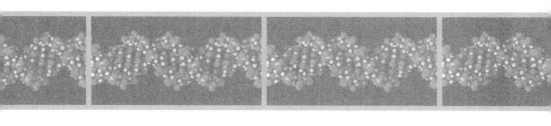

Preface

This book, *The Purposeful Universe,* is meant to serve two somewhat different purposes. On the one hand, this book presents a theory to be scrutinized and critiqued by the professional scientific community. A proposal for a new theory of biological evolution as an alternative to Darwinism requires that supporting evidence from many different fields be provided in some detail. On the other hand, this theory is intended for seekers of the truth who do not have any formal scientific training. Since the substance of this theory addresses some of the most significant questions all human beings raise at one point or another in their lives—such as *Who am I?, Why am I here?,* and *Where do I come from?*—I also hope that the information imparted will serve a broader audience.

This material has not been published in peer-reviewed articles because the theory needs to be presented in its entirety for the sake of clarity. If parts of it are removed from the overall context, this theory would probably appear so completely outside the ruling paradigm of science that no peer reviewer would be able to understand it. As I hope to demonstrate, a theory about biological evolution that holds together, and does not in the final analysis look upon life as an accident, requires a very broad perspective, one that includes Mayan cosmology as an indispensable part. This new theory about the evolution of the universe in fact spans so many fields that no professional scientist, and I am no exception to this myself, can have firsthand knowledge of all the fields that need to be covered for its development. This also means that many well-read nonprofessional scientists are as well equipped to study and evaluate this theory as someone who has a scientific degree in just one of its specialized fields. Consequently, I think this book should be accessible to a fairly broad audience. Meaningful science is never overly complicated, and a true theory should be clear enough that it is accessible to just about everyone with a strong interest. I think we have now come to a point where the closing words of Stephen Hawking in *A Brief History of Time* may apply:

> If, however, we were to discover such a complete theory, its general principles should be so simple as to be understood by everyone, not just a few scientists. Then we shall all, philosophers, scientists, and just ordinary people, be able to take part in the discussion of the question of why it is that we and the universe exist. If we find the answer to that, it would be the ultimate triumph of human reason— for then we would know the mind of God.[1]

It is exactly this kind of discussion that I would like to see result from the study of the theory presented, and in consequence I have decided to leave it up to *all* readers to decide for themselves, or among themselves, whether this new theory of evolution makes sense and is useful for them in pondering the big questions of life.

I should point out that this book contains the *original* theory; this is not a book popularizing a theory that has been developed somewhere else. Despite the amount of detail that is included, I think that generally the arguments are easy to follow and the math is elementary. I hope with this book to directly share the excitement of the development of a new scientific theory and the methodical detective work that is needed to verify it. I also encourage a critical reading when it comes to the facts and logic presented. While intuition is very important for developing a new theory, I feel that intuition by itself is of little value if it cannot be verified by facts and logic.

To some, the title of the book may seem controversial and its focus, biological evolution, even more so, as this has been the center of a hard-fought battle between religious creationism and Darwinism for 150 years, a struggle that has probably even intensified in the past decade. The purpose of this book is to propose a resolution to this conflict through a theory that brings our understanding to a higher level from which our perspective can be unified. As it turns out, with this theory both sides in the protracted conflict actually get to be right in very fundamental ways, which incidentally may be completely unsuspected even by their protagonists. Yet to arrive at such a unified model, it is necessary to criticize earlier ideas that by necessity serve as our points of departure. Thus I spend some time explaining the theory of Darwinism in its current form, which is what dominates the scientific environment, along with the big bang theory, which explains the beginning of the universe. These make up the creation story that is embraced by the modern educational system. It is therefore an urgent matter to find out if these theories are true or if the facts that they are based on can more favorably be looked upon in an entirely different way. I spend much less time criticizing Young Earth Creationism, not because I embrace it, but because it ignores empirical evidence to an extent that such a discussion would not seem meaningful.

Hence, the present book is primarily a scientific theory about the basic nature of the universe and its evolution, especially pertaining to physics and biology, and so it deals mainly with the history of the cosmos—where we come from. It does not directly address the question of what will happen in the years and months leading up to 2011 and after—where we are going. My understanding of the history of humankind and its future has however already been discussed quite extensively in two previous books: *Solving the Greatest Mystery of Our Time: The Mayan Calendar* and *The Mayan Calendar and the Transformation of Consciousness.*[2] These two books complement the present one. Nonetheless, I think that this new book is instrumental for readers who want to understand the current situation of humanity. For me, a discussion of the future of humanity based on the Mayan calendar has always required proof that this calendar is accurate, that it can be used to match events in history or biology with the passage of time. I do feel the present book adds crucial evidence that the true Mayan calendar, with nine Underworlds and thirteen Heavens, is useful for understanding evolution, and I believe this is necessary for a meaningful understanding also of the future. Especially when the media are defering all their interest on the meaningless date of December 21, 2012, it seems all the more important to provide information concerning what the true Mayan calendar tells us about ourselves and the purpose of the universe.

In a sense this book is the third in a trilogy that with the two mentioned above develops a complete theory about the evolution of the universe. While it might have seemed natural to start this trilogy with the evolution of the physical universe and biological species and move on to human history, the books have if anything been going in the opposite direction. Hence, this trilogy ends with the current book, with its more basic analysis of the nature of the universe. The main reason that the scientific book about physical biology comes last is that it is only in light of some recent discoveries in the scientific literature that it seems possible to develop a coherent theory of evolution without large gaps and unfounded assumptions. This third book is in my own mind the deepest and most powerful for understanding the nature of the purposeful universe that we are part of, and hopefully it will also help readers to complete their own pictures of where they come from and where they are meant to be going.

The Purposeful Universe has nine chapters, and I would like to briefly outline the contents of these here. The three first chapters are all essentially introductory ones presenting background on universal cosmology, the nature of time, and biological evolution. They highlight not only the current state of knowledge and the positions of some schools of thinking in these areas, but also a few findings that are especially important for a new start. Chapter 4 presents results that favor a new theory of evolution, the Tree of Life theory, which is the central focus of this book. It corresponds to the "Results" section of a scientific article and presents the underlying facts and data that support the new theory of biological evolution. Chapter 5 offers information on physical cosmology, combined with a new interpretation of quantum theory consistent with the view that we are indeed living in a universe that is intelligently designed and fine-tuned for life. Chapters 6, 7, and 8 present different aspects of the new theory of biological evolution. Chapter 6 discusses why the universe evolves in the first place as a function of the shifting quantum states of the Cosmic Tree of Life and shows how lower galactic and planetary levels evolve because of their entanglement with this phenomenon. It focuses on the place of biological evolution in the larger context of geology and astronomy. Chapter 7 explains the origin and evolution of life on the biochemical level, while the actual cellular mechanism of biological evolution is described in chapter 8. Chapter 9 discusses some of the wider existential ramifications of the new theory of biological evolution. These took me somewhat by surprise, and the consequences have turned out to be much more far-reaching and exciting than I had anticipated when I started writing this book.

This is not a self-help book that tells anyone his or her own personal purpose or makes recommendations about how to live their lives. Yet I think it is impossible for anyone to come to clarity about their individual purposes without knowing that the universe we live in is purposeful and has a direction that our own lives may be part of. The existence of purposefulness to the universe has been denied by modern science, and in demonstrating this purpose, it has been necessary to break some of its rules. Instead of saying that the universe lacks a direction, which is the current scientific norm, I advocate a return to the philosophy of some of science's early pioneers, who did not avoid addressing the great questions of life and its origin. Yet, for the very reason that I have broken certain taboos, I feel the theory has become much more encompassing and meaningful than it would otherwise. It now tells us more about the beauty and intelligence of this creation and will hopefully allow more people to experience the beauty of the process of scientific discovery as well.

1

The Universe Is Not Homogeneous

And It Never Was!

Pour atteindre à la vérité, il faut une fois dans sa vie se défaire de toutes les opinions que l'on a reçues et reconstruire de nouveau, et dès le fondement, tous les systèmes de ses connaisances. (To arrive at the truth, once in your life you have to rid yourself of all the opinions that you have received and reconstruct anew, from the foundation, all the systems of your knowledge.)

RENÉ DESCARTES

THE OLD BIG BANG THEORY AND THE COSMOLOGICAL PRINCIPLE

Since the 1960s there has been a relatively strong consensus within the scientific community that our universe was born in a "big bang," the description of which is essentially as follows: First there was

1

nothing. Then, some 15 ± 5 billion years ago, a point of extreme density started to expand, growing into a fireball, which for the first 10^{-43} seconds was so hot that energy and matter could not be distinguished. Before this very early point in time, the four basic forces of physics (gravity, electromagnetism, and the strong and weak nuclear forces), had been unified in one force, but then (at 10^{-43} seconds) the different forces of nature started to separate. After 10^{-35} seconds an inflationary phase started in this expansion, during which the space of the universe increased a trillion billion times faster than initially. Although there is a consensus about the existence of such an inflationary phase that lasted until 10^{-32} seconds after the very beginning, no one has provided an explanation as to why this phase began or why it came to an end. The existence of such an inflationary phase, however, has to be assumed to explain the current state of the universe. When the inflationary phase was over, elementary particles such as electrons materialized, and the temperature of the universe started to cool down so that after 34 minutes it was down to 300 million degrees K.[1] Over the next 300,000 years, the purportedly homogenous universe that resulted from this unfathomably large explosion and expansion, called the big bang, cooled down even further, and the simplest of the atoms formed, hydrogen, helium, and lithium. From those cooler temperatures came an afterglow of cosmic microwave background radiation (CMB) at a temperature of only a few degrees above absolute zero.

This scenario for the beginning of the universe, or one with much more detail, is consistent with many observations. Despite this apparent success, the problem remains for cosmologists to explain how a heterogeneous universe with the kind of order we know today from our everyday lives could emerge from such an enormous explosion. After all, the universe we see around us is structured and does not look like the remains of a fireball of billions of degrees. If nothing else, we know that the universe is structured into billions of galaxies, and life on our own planet looks very orderly as well, and it would seem that some kind of ordering principle must have existed to account for this. Instead, to explain this transition from homogeneity to an ordered heterogeneity, physicists have referred to what they have called random fluctuations in the density distribution of the early universe. These, physicists have proposed, created higher concentrations of matter in certain locations until they have started to accrete into galaxies, and within them star systems later supposedly emerged also because of random density fluctuations. The scientific theory about the origin of the universe has thus been dominated by a philosophy of randomness.

The reason you are here to read this, in other words, according to this philosophy of randomness, is because after a giant explosion, undirected random fluctuations started to generate material structures such as galaxies and stars, which in turn gave rise to other processes, purportedly generated by random mutations, from which all organisms, including yourself as a human being, are believed to have emerged. And so, in this philosophy of modern science, the current apparent orderliness of the universe ultimately is viewed as a result of a high number of accidental events. Given the total absence of purpose and the predominance of random processes in this cosmological scenario describing our origins, we have no reason to be surprised that Nobel Prize winner Steven Weinberg, in his much-acclaimed description of the origin of the universe, *The First Three Minutes,* concluded: "The more the universe seems comprehensible, the more it also seems pointless."[2]

Why then has the view of our universe presented by modern physics come to seem so pointless? This is not a very difficult question to answer. Modern science simply rejects from the outset any idea that there could be a purpose, or intelligent design, in the universe as "unscientific." If you try to sneak in a comment in a scientific article that implies that there may be some kind of purpose to the universe, you can be certain that it will be rejected by the journal. Because any implication of a purpose has been censored out of thousands of articles, it seems obvious that more broadly published works, like Weinberg's, will conclude that science does not see any higher purpose anywhere in the universe. It is not very surprising that if science acknowledges only assumptions that exclude any higher meaning of life, this will also be how it describes the nature of the universe. Obviously then the view of the universe that science has offered is that it is pointless. While this view may not be shared by all, or even many, individual scientists, this is nonetheless the officially endorsed view, and on this basis some scientists will even attack those who suggest that the universe was created by a higher intelligence. In my own view, by excluding purpose from any inquiry at the outset, science has in fact put a lid on the search for the truth. This philosophical foundation of randomness and lack of purpose also deviates strongly from that of some of the towering pioneers of science, such as Descartes, Kepler, Newton, and Einstein.

Yet, as we know, it is unavoidable that ideas in science undergo change as a result of new discoveries, and I feel it is becoming increasingly obvious that the picture that science has presented of our origins has significant gaps, such as the underlying reasons for the inflationary phase or the density fluctuations mentioned above, to name just two. Personally, I think that there are too many unexplained processes or phenomena in the model of the origin of the universe that science currently presents, and that it is necessary to deeply question the basic assumptions. Only by doing so can we arrive at a theoretical basis that has a higher explanatory power than the present paradigm of randomness. If purpose and meaning are excluded from scientific theories about the origin of our universe, these will simply continue into a dead end and increasingly alienate people at large, who seek a meaningful understanding of the universe from the process and results of science. A new significant discovery that shows that the universe was never disorderly to begin with may then aid us in providing a new context for our origin.

THE DISCOVERY OF THE "AXIS OF EVIL"

The application of many of the theories of modern physics to cosmology has been based on what has been called the cosmological principle. The assumptions of this principle are that the universe at the largest scale is homogeneous and isotropic (meaning that it looks essentially the same in all directions). The assumption that this principle is valid limits the number of possible models of the universe that may be proposed and excludes, for instance, a nested model in which some of its structures, such as galaxies and planets, have a certain degree of autonomy. Use of the cosmological principle has however simplified development of mathematical models proposed to describe the overall structure and behavior of the universe. The theories of special and general relativity for instance are based on this principle and have been assumed to apply uniformly everywhere. The cosmological principle has often been presented like a proven fact rather than an assumption. Data are now accumulating however that indicate that this principle, basic to the applicability of unified field theories of physics, may not hold true. I believe that this means that we are in for a real revolution in cosmology, a revolution that is more than a paradigm shift and may come to entirely redefine what science is about.

For some time now it has been realized that the study of cosmic background radiation, or CMB for short, may serve to probe the relevance of the cosmological principle. One of the predictions from the big bang theory was that the high temperatures generated at the sudden early expansion of the universe would give rise to an afterglow in the form of CMB at a few degrees above absolute zero. The detection and measurement of CMB exactly in accordance with this expectation was a very clear verification of the big bang theory, and for this feat its discoverers, Arno Penzias and Robert Wilson, shared a Nobel Prize in Physics in 1978. The study of CMB also became the basis for projects designed to test the relevance of the assumptions of homogeneity and isotropy in the universe. If these assumptions were true, so it was argued, the temperature distribution would not display some clearly discernible directional pattern in the early universe some 15 billion years ago, and the cosmological principle could hence be considered as valid. Data from the initial satellite launched in 1989 to measure the CMB indeed only discovered minor anisotropies (irregularities) in the background radiation, which were interpreted as indicative of randomly distributed early density fluctuations[3] that later would give rise to galaxies. Hence, no large-scale structures were discovered in these initial studies and the CMB was still consistent with the cosmological principle.

In 2003, new satellite measurements recorded the CMB by means of the so-called Wilkinson Microwave Anisotropy Probe (WMAP),[4] which provided more accurate data from the afterglow. When this data set was analyzed mathematically, an axis and polarized fields of temperature were discovered by Swedish-American cosmologist Max Tegmark and coworkers.[5, 6] These fields were organized somewhat like the panels of a basketball, which defined the direction of an axis through the early universe (figure 1.1). Kate Land and Joao Magueijo of Oxford University later verified these findings in an article entitled "The Axis of Evil."[7, 8] This unfortunate label, with associations to President George W. Bush's foreign policy, has remained attached partly because the label is catchy and partly because the finding seems so disturbing to the foundations of established cosmology. (I prefer the term Central Axis.) As it now appears from the WMAP, the cosmos is really organized somewhat like a meatball on a toothpick, where the center of the meatball, really then what seems to be the center of the universe, is located in the direction of the constellation Virgo.[9] Curiously, the associated Central Axis, which exists on an unfathomably large supergalactic scale, lies in the plane of the ecliptic and has a direction parallel to the equinoxes of our own Earth's planetary orbit.[10] To give a sense of the scale of this axis we may note that our own galaxy, the Milky Way, has a diameter of about 100,000 light-years, while the diameter of the visible universe is around 30 billion light-years. Thus, we are talking about an axis that may be

Figure 1.1. The cosmic microwave background with the "Axis of Evil." Top: The cosmic microwave background in the original image from the WMAP study. Middle: After mathematical processing of the temperature variation, a structure around the quadrupole axis becomes evident. Bottom: An octupole axis aligned with the quadrupole was discovered, which demonstrates the existence of an "Axis of Evil." Source: NASA and the WMAP team. With permission from A. de Oliveira-Costa et al., "The Significance of the Largest Scale CMB Fluctuations in WMAP," Phys. Rev. D 69 (2004): 063516.

roughly a million times the size of our own Milky Way galaxy, in which our own little planet is a mere dot to begin with.

We should be aware that these measurements of the CMB give a picture of what the universe looked like when it was only 300,000 years old, corresponding to 0.002 percent of its present age. (This picture of the early universe is thus comparable to a photograph of a human infant twelve hours after it was born.) Hence, the CMB picture shows the temperature variation in the universe essentially as it appeared at the big bang. Whatever structures that are apparent in this picture are inherent in our universe from its very inception. It is for this reason that these CMB studies have such a fundamental importance for our understanding of the origin and evolution of the universe, and they indicate that the universe has always contained a potentially organizing structure.

Naturally, when so much is at stake regarding the basic assumptions and future of modern physics, the discovery of a Central Axis of the universe has been questioned.[11] Studies of the axis based on other sets of data have however supported the validity of its existence in dramatic ways: First, it was found that the polarization of light from quasars (some of the brightest and most massive astronomical objects known) was influenced by their proximity to this Central Axis.[12] An interesting twist of this particular study is that the polarization *corkscrewed* around the axis, and its authors suggested that a potential explanation of this effect is that the entire universe rotates around this Central Axis. This would then be consistent with the idea that *the universe as a whole is a spinning vortex generated by the axis and that it emerged as such from the very beginning.* A second, equally dramatic, finding was made by Professor Michael Longo at the University of Michigan, who studied the handedness of spiral galaxies (whether they revolved clockwise or counterclockwise) throughout the universe. He found that a line separating the preference for the two types of handedness approximately lined up with the axis previously discovered in the WMAP (figure 1.1).[13] The axis that our own particular galaxy revolves around, as well as those of most other galaxies observed, was also found to be aligned with the Central Axis.

Because of these recent findings, the Central Axis is a really exciting topic. The study of the handedness of galaxies was made independently of the WMAP study and could not have been distorted by some unknown influence from our own Earth or galaxy. Longo's study was based on a few thousand galaxies, which would normally be considered a large enough sample, and he concluded: "A well-defined axis for the universe on a scale of ~170 Mpc would mean a small, but significant, violation of the Cosmological Principle and of Lorentz symmetry and thus the underpinnings of special and general relativity."[14] Because of the serious implications of such conclusions for the future of cosmology, Kate Land made a call to the public to participate in an expanded study that measures the handedness of a large number of galaxies. Regarding the "Axis of Evil" she stated: "It's a massive and alarming claim, which if true, forces us to come up with a new framework for cosmology."[15]

Dr. Land's claim is no exaggeration. What science has hit upon here is, I believe, one of the most consequential discoveries of all time, something that will come to profoundly change how we look upon the universe and its purpose. What I am suggesting is that the Central Axis is the fundamental space-time organizer of our universe. Its discovery, as we shall see throughout this book, may point out a way to recognize the universe as purposeful and endowed with a creative intelligence that in ancient times were taken for granted by most peoples on this planet. This, in turn, would mean that we for the first time would have some credible explanation as to why the universe has turned out to be organized in an orderly way. The first attempt at repeating Longo's results of a polarized handedness of galaxies however was reported as negative,[16] although interpretations of this new study diverge,[17] meaning that more studies are under way to clarify this particular issue. Regardless, the Central Axis still stands, and if Longo is right, his data point to a universe that was polarized from the very beginning with regard to a handedness of galaxies.

What is more, if the handedness and spin axes of galaxies are directly related to the Central Axis of the universe, this means that the formation of galaxies may not just be a result of random fluctuations causing them to rotate. The spin of the galaxies would instead be related to that of the Central Axis and probably emerged in resonance with this. This would mean that all the galaxies of the universe are connected to this Central Axis, and also possibly to one another, in a form of entanglement at the largest scale possible. If the different galaxies of the universe are not independent, but retain a connection to the overall polarized structure created by the Central Axis, this would also favor the idea that their evolution is connected and synchronized, something that we will see later is crucial for the evolution of life.

I am aware that these findings may initially seem very remote from the present everyday existence of most people, but as we shall gradually see, they will prove to have a very fundamental meaning for who we are. They create an opening toward an entirely new worldview, one where the world we live in is not seen as a result of random events and is far less fragmented and confusing than it currently seems. It is also clear from the quotes above that the findings may turn out to be very consequential for the dominating theories of physics. The theories of special and general relativity, formulated by Albert Einstein in the early twentieth century, have for almost one hundred years provided the general framework for cosmology. A crucial idea that relativity theory is based upon however is that a preferred system of reference for observing the universe does not exist. Relativity theory makes cosmological studies independent of the movement and positions of different observers. Since the Central Axis of the cosmos violates the cosmological principle and potentially presents a preferred system of reference, the new discoveries directly threaten the underpinnings of general and special relativity and other unified field theories of physics. It is too early to tell if this means that Einstein's equations will be invalidated, but it does mean that there are now reasons to question relativity theory as the fundamental theory of space-time.

The core discovery however is that the existence of the Central Axis shows that there was structure in the universe from its very inception, which puts the whole randomness philosophy and purported purposelessness of our universe in question. This by itself allows for the development of cosmological models in which the universe is not pointless when it comes to how life, consciousness, and intelligence have evolved. To make sense of the new discoveries and acknowledge the purposefulness of the universe it will however serve us to go back to some of the earlier traditions of humankind to pick up the necessary pieces of the puzzle of the universe that have been lost along the way. This pertains especially to the cosmological tradition of the Maya. In scientific terms what I propose here is that the Central Axis is a preferred system of reference and that a fundamental space-time theory of the universe needs to be based on this very axis in order for an evolutionary theory to make sense.

THE NEW BIG BANG THEORY AND HUNAB-KU

The "Axis of Evil" of Modern Science Is the "Tree of Life" of the Ancients

The hypothesis that will be developed in this book is that the Central Axis that modern cosmologists have now uncovered is the Cosmic World Tree, the Hunab-Ku of Mayan cosmology—which in the Jewish tradition is known as the Tree of Life—and that all of evolution emanates from this. In Viking mythology a World Tree, Yggdrasil, was seen as the center of the universe, which connected its nine worlds. In Mayan language, *Hunab-Ku* literally means the "One Lord" (Hun Ahau), who is mostly described as the One Giver of Movement and Measure or the One Giver of Energy and Boundaries (figure 1.2). As is apparent from the Mayan Tree of Life symbol shown here, its most marked features are a separation between light and darkness; a yin/yang polarity; and a spiraling, or corkscrewing, movement. That such a center of the cosmos indeed exists is described by Mayan cosmology, and I have elaborated on this in both of my previous books,[18] although at the time of their writing there was no empirical evidence at hand to demonstrate the existence of such a Central Axis.

I propose that the evolution of life and consciousness can be understood if we look at the universe as a hierarchical organization of Halos

Figure 1.2. Hunab-Ku, or the One Lord, is the creative center of the universe in Mayan cosmology. This figure also goes by the name of the Eight-Partition-Place, the One Giver of Movement and Measure, and the One Giver of Energy and Boundaries. The symbol incorporates certain traits of the creative center of the universe such as the Tree of Life, a spiraling movement, and the yin/yang polarity. Courtesy of José Argüelles.

(generated by spinning "Trees of Life"), ranging from the level of the universe in its totality down to the elementary particles. I am intentionally retaining the term *Tree of Life* for the central Cosmic Axis, although this is laden with mythological meaning and may sound archaic. This is mainly because the phenomenon could easily be trivialized, or seen as merely material, if it is given a name from the terminology of modern physics. The use of the term *Tree of Life* reflects the knowledge that the ancients had of this axis, a concept of an unfathomably large "tree," presumably based on inner visions, that created life in all of its aspects. This Tree we may get an idea about from old Mayan representations (figure 1.3), and one way of visualizing the Tree of Life is that from its center, the so-called *Yaxkin,* three perpendicular axes emerged that through their vibrations created a universe in three dimensions. The Tree of Life also created the four directions locally on Earth from the polar axis that were reflections of the three dimensions of length, width, and depth on the cosmic level. Its three axes are thus the "boundaries" or "measures" provided by Hunab-Ku. The Tree of Life also created periods of time as well as the human beings and all animals and plants of the world. Hence, it was often portrayed with representations of maize and human heads coming out of its branches (see figure 1.3), and it was seen as the central creative principle of the universe, the creator of all life. As a matter of terminology I will refer to the Central Axis of Hunab-Ku (figure 1.2) as the Cosmic Tree of Life and the sphere that surrounds its polarized field as its Halo, which is dominated by a yin/yang polarity.

What makes the suggestion that the Central Axis of the universe is the Tree of Life of the Maya scientifically meaningful is, as we shall see, that *it is a testable hypothesis and not a statement of faith.* The reason for this is that the Mayan calendar provides a description of the rhythms of creative energies emanating from this hypothetical Tree of Life and so it is testable to what extent the actual rhythms of evolution conform to the time plan that this calendar specifies. Thus, although we are testing the hypothesis that the Central Axis is a manifestation of the Mayan One Lord, which may seem like anathema to many professional

Figure 1.3. Mayan Tree of Life from the Temple of the Foliated Cross in Palenque. In the Mayan view this World Tree was the creator of all life, including the human beings whose heads are shown in its branches. Courtesy of Linda Schele and David Freidel.

scientists, we will do so by means of empirical science. If the evolutionary processes can be convincingly shown to adhere to the rhythms of the Mayan calendar, we will have to conclude that they emanate from the now discovered Central Axis of the universe and that this is indeed what the ancients called the Tree of Life. If this is true then the Tree of Life is not, as sometimes has been thought in recent centuries, merely a symbolic or mythological concept, but something very real that has enormous consequences for our view of the origin of the world and of ourselves.

Even if it may to some initially sound outrageous that the Tree of Life actually is real, there is something that speaks very strongly in favor of such an existence. This is that the notion of a World Tree, sometimes with a snake or a dragon whirling up and down it (maybe related to the corkscrewing of the polarized light from nearby quasars mentioned above), in ancient times was the most widespread myth among the peoples of the Earth. We know of the existence of this myth from the Norse to the Hindus and from the Maya to the Maori. It is part of the lore of the San people of the Kalahari desert, arguably the oldest existing culture on Earth. For Westerners the most well-known description of the World Tree is however the Tree of Life in the Jewish creation story in Genesis. The Tree of Life plays an equally important role as the very foundation of creation in the Icelandic Sagas and the Popol-Vuh of the Maya. Sometimes, special trees (actual living green trees) serve as symbols for the Tree of Life, like the kauri tree of New Zeeland, the ash in Scandinavia, or the ceiba in Guatemala. In other cases the representations of the World Tree are more symbolic and reflective of spiritual energies, such as the Kabbalah tree, the medicine wheel of American Indians, or Svetovid of the Slavic peoples. It is clear from the myths of the Maya and the Vikings that the Tree of Life was believed to exist on an immense cosmic scale. If the Tree of Life was a mere superstition, or an abstract symbol without any real manifestation, it is difficult to understand why the idea of its existence would have become so widespread among so many, and widely different, cultures of our planet.

In Mayan cosmology, Hunab-Ku imposes polarized fields on the universe (figure 1.2), creates coherent organizations of life on different levels, and maintains their synchrony according to the principle of "as above, so below." In the hierarchical organization of life that Hunab-Ku creates, the boundaries and energies of the microcosms are ultimately defined by the macrocosms in a way that, in line with the great thinker Arthur Koestler, could be called holonomic. This means that the universe is formed as a nested hierarchy of Holons, structures that are whole in themselves and yet interconnected. The model that will be developed here also borrows from the holographic model of David Bohm and Karl Pribram, according to which the universe we perceive is a product of interference patterns of wave forms that provide an implicate order for the cosmos.[19] What to me however seems to be missing for the holographic model to become tangible is an understanding of the source of these wave forms and how they may have generated an orderly universe. What I will be proposing here is that the source of these wave forms is the Central Axis, or Cosmic Tree of Life, which creates a nested hierarchy of Holons, or organizations of life that are microcosmic reflections of the Central Axis.

To avoid confounding the present theory with earlier models that I have not fully described, I prefer to talk about the universe as Halographic, since there is much data to indicate that the Holons are formed by Halos generated by spinning Trees of Life at different levels of the universe. In this model the higher macrocosmic levels of life in the universe, created by the Trees of Life on their respective levels, are senior to the microcosmic levels. The latter are however not merely identical copies of the former, reduced in size. Each lower level instead defines a distinct level of organization with unique characteristics within the framework provided by a higher level. The origin and evolution of life cannot in this model be understood by studying it only from below, from an understanding of molecules, electrons, or strings—in other words, by a reductionist approach. Ultimately we need an understanding of the highest, all-encompassing, cosmic level of the Tree of Life that appeared at the big bang.

In the view of the ancient Maya, the Tree of Life, or Heart of Heaven, was the creator of everything (even though it may have needed to be activated by Hunab-Ku, or the One God, in the beyond). Before time existed, "there was only the incomprehensible divinity of Hunab Ku, permeating the Heart of Heaven, which slumbered for seven eternities. Then by the power of his word Hunab Ku thrilled the Heart of Heaven."[20] Considering that a Central Axis has now been discovered, which existed from the very beginning of the universe, it is warranted to propose a new big bang theory consistent with this creation story. In this, the Tree of Life is not seen as a manifestation of the big bang. Instead, *the big bang was a manifestation of the creation emanating from the Tree of Life.* Thus I will speak about the Tree of Life as a *Platonic form,* which means that it has a geometric existence that is prior to and beyond any physical manifestation.

The fact that the universe seems to have had a Tree of Life structure from the outset is clearly consistent with the interpretation that our universe was designed by an intelligence. Although to mention this may mean breaking a taboo in science, there is nothing in the physics of the big bang theory that contradicts this. The Central Axis will here be seen as a creative source that introduces yin/yang polarities of creation, and its identification as the Cosmic Tree of Life will have wide-ranging ramifications for how we may understand not only the big bang, but even more so the continued evolution of life in the universe. What I proposed with this new big bang theory is that as the Yaxkin, the center of the Cosmic Tree of Life, grew to a certain size in the original fireball, it started to separate out the four forces of nature and created a polarization that pervaded the cosmos (which later, when the galaxies started to develop, presumably determined their handedness). After 10^{-35} seconds, three axes emerged from the Yaxkin, which gave rise to the inflationary phase that expanded the volume of the universe some 10^{100} times, or, in other words, the three axes developed from the central point into a creative coordinate system, including the Central Axis, so that the three basic dimensions of space came into existence.

In a rapid sequential expansion, Tree of Life Halos at lower levels of organization, also with three-dimensional coordinate systems, were generated throughout the cosmos and contributed to this inflationary phase through their own expansion (figure 1.4). These Halos at different organizational levels emerged in a certain specific order, with those at a lower level inside those at a higher, thus the Cosmic Tree of Life initially gave rise to seeds of Galactic Trees of Life that developed into galactic coordinate systems and provided the primordial coordinate system for the later development of galactic vortices. Within each Galactic Tree of Life, in turn, Star System Trees of Life were seeded, which are subordinated to, and in resonance with, the galactic level, thus ensuring that the star systems are integrated into the organization of the galaxy that seeded them. In this way, seeds are planted for Trees of Life at lower levels of organization, which are always subordinated to their next higher holonomic levels. This ensures coherence, ranging from star systems to planets, organisms, cells, and elementary particles, according

Figure 1.4. The emergence of Tree of Life Halos at different levels in the big bang. According to the new big bang theory presented here, this was primarily an organizing event in which the Cosmic Tree of Life emerged as a three-dimensional coordinate system surrounded by a Halo. This initiated an inflationary phase of expansion resulting in a nested hierarchy of spinning Trees of Life with Halos at different organizational levels, such as galaxies, stars, and planets. Courtesy of Bengt Sundin.

to a *seniority rule*—as above, so below—where microcosms develop within the framework of macrocosms. Ultimately the Trees of Life in this model are all subordinated to the Cosmic Tree of Life at the largest scale of the universe. When the inflationary phase of the universe was over at 10^{-32} seconds, the universe was filled with electrons at one of the lowest levels of this chain of unfolding Trees of Life. The inflationary phase of the universe was completed when the dimensions of the universe and the seeds of lower-level Trees of Life had all been introduced in an organized way. From its very largest scale to the smallest, the different levels of the universe are connected through the Halographic resonance of these creative coordinate systems, a setup that allows for its coordinated evolution and overall coherence, and does not hinge on any precarious random processes.

In this model, it is the Cosmic Tree of Life that gives our universe its three dimensions of space, something whose origin has not been explained in any previous big bang theory. It is not trivial or in any sense self-evident that our universe has three dimensions of space, since in principle the primordial fireball could have developed into any number of dimensions, and theories proposing exotic numbers of dimensions have been proposed. Yet such theories are inconsistent with our everyday experience as human beings, in which it is three dimensions of space that we really experience, and it may be that it is only in a three-dimensional universe that life can emerge. This explanation was suggested in 1955 by Whitrow, who among other things put forth that three dimensions of space are a necessary prerequisite for a chemistry-based life.[21] More recently, Martin Rees has in *Just Six Numbers* made a renewed case that three dimensions are necessary for the emergence of life, since in a universe with more than three dimensions of space, planetary orbits would be destabilized. Against a universe with two or one dimensions he argues: "It is impossible to have a complicated network without wires crossing; nor can an object have a channel through it (a digestive tract, for instance) without dividing into two. And the scope is still more constricted in a one-dimensional 'lineland.'"[22] It seems, in other words, as if three dimensions of space are a necessary prerequisite for our existence. To my knowledge the Tree of Life hypothesis is the only one that has ever been proposed to explain why the universe has exactly this number of dimensions.

In conventional science, quantum events are limited to the micro world. Here we will however look upon the emergence of the three-dimensional Cosmic Tree of Life as the primordial quantum event. To say that this was a quantum event means only that it happened without intermediate states. As all quantum events it just happened and resulted in a discrete state of the Cosmic Tree of Life and the lower levels entangled with it. In chapter 6, we will develop this vision of the cosmic origin more fully, and evidence will be provided that quantum events occur in the universe also at a macro scale.

At the end of the inflationary phase, 10^{-32} seconds after the big bang, there were, of course, no ready galaxies, not to mention planets, animals, or plants, but in the Tree of Life hypothesis the seeds for the emergence of all these things had been planted in the lower-level Halos surrounding lesser Trees of Life. In this view, the evolution of life begins with the big bang, rather than suddenly popping up by accident 3.8 billion years ago on Earth. Life was seeded from the very beginning, when the life-generating Cosmic Tree of Life created an expansion of low-level Halos throughout the universe. In the view presented here, the universe emerged with the very purpose of generating life, and this purpose started to manifest from its inception. It is this evolution of life that, as we will see, takes place according to the rhythm provided by the Mayan calendar. It should now be pointed out that the Mayan calendar is not an astronomical calendar in the usual sense of the word, but is rather a reflection of creative pulses emanating from the Tree of Life. The Mayan calendar is thus an expression of the time factor in creation and evolution. Not only space, but also time was generated by the Tree of Life, the One Giver of Energy and Boundaries, which in my new theory is the fundamental source of space-time. In this new scenario the big bang is not an explosion in which matter emerges out of nothing in random ways. Rather it is an expansive, or inflationary, *stage-setting organizational event,* in which the Cosmic Tree of Life and its Halos define the spins at different levels of the universe, making the energy already existing in the quantum vacuum materialize as matter in the form of particles and objects with clearly defined properties. The existence of such an underlying energy field is fairly universally recognized by physicists.[23] Here we will focus on the evolution of the physical reality that has emerged from it.

This new big bang scenario—unlike the old—can explain why a Central Axis has pervaded the universe from its very outset, as well as why an inflationary phase in three dimensions took place shortly after the beginning. Unlike the old theory, the new big bang theory explains how an ordered heterogeneous universe came into existence from a hierarchical organization of Halos, or wave forms, generating the seeds of structures from galaxies down to electrons. It also provides a background to the many observations that the universe is fine-tuned for the emergence of life.

THE PHYSICAL CONSTANTS OF THE UNIVERSE
Fine-tuned for the Generation of Life

The time points I have used above in the new big bang scenario have been taken straight from the old big bang scenario and are based on estimates of the original rate of expansion of the universe. Like the three-dimensionality of the universe, this rate of expansion, and its relationship to the opposing force of gravitation, is one of many examples showing that the basic features of this universe seem to have been remarkably fine-tuned for the generation of life from the very outset. If this initial rate of expansion had been only slightly smaller, the universe would have collapsed back on itself shortly after the big bang. Stephen Hawking has estimated that if the rate of expansion (at the time when the universe was 10^{10} °K) had been just one part in a trillion less than what it actually was, the universe would have collapsed on itself a long time ago.[24] If, on the other hand, the rate of expansion had been higher, matter would have been so disperse that galaxies, solar systems, or planets would never have been able to form. Either way, no life would have emerged in the universe.

Thus the initial rate of expansion of the universe, presumably determined by the rate of expansion of the Cosmic Tree of Life, is a critical factor that the emergence of life hinges on. Other basic constants of nature also have been critical for defining the way our universe looks and has developed, and these have been discussed by Martin Rees.[25] Some of these constants express the strengths of four basic forces of nature, such as α_{EM}, (the electromagnetic interaction), α_G (gravitation), α_w (the weak force), and α_s (the strong force). The emergence of life in the universe is directly dependent on these constants being within a very narrow range of the values that they actually have. For those who believe that life is an accident, it has obviously presented a challenge to explain the preciseness of these values, and so far no unified theory has been provided to account for this.

For instance, if the constant, α_s, the strong force, which causes the attraction between different protons and neutrons in the nuclei of atoms, had the value 0.006 rather than 0.007, then a proton would not be bonded strongly enough to a neutron. In a universe with the lower of these constants, helium and all the elements necessary for life, including carbon, oxygen, and nitrogen, could not be formed.[26] The only type of atom that would exist in such a hypothetical universe is hydrogen (which has only one proton in its atomic nucleus), and since in such a universe hydrogen would not be able to fuse to generate helium, there would be no stars generating heat and light, which as a consequence would preclude the emergence of life. If in another hypothetical universe the constant was higher, 0.008, two protons would attract one another so strongly that all hydrogen would rapidly form higher elements and thus disappear at such an early point that there would not remain any fuel for the stars in this universe either. All carbon would rapidly burn to oxygen and then go on to create metals at a rate that would preclude life.

In another example, if the number α_G/α_{EM} (the relative strengths of gravity and electromagnetism in the universe) was a little bit higher, galaxies would form much more readily and stars would have much shorter distances from one another than in our current universe. As a result, encounters between different star systems in the galaxy would be much more frequent and create constant disturbances on the planetary orbits, and so pose grave threats to life everywhere. What is more, the strong gravity in such a universe would give rise to stars with much larger masses than is now the case, and as a consequence they would have much shorter life spans, maybe on average ten thousand years rather than about ten billion years, which is estimated for our own Sun. Such shorter lifetimes of stars would likely not provide enough time for the evolution of life. For our existence we are very dependent on having a strength of the force of gravitation that is very close to what it actually is in our universe.

One might stand in awe of the exactness of any single one of these constants of nature and how they seem to have been perfectly attuned for the generation of life. Our existence has in a very real sense narrowly depended on a set of basic constants of nature having close to exactly the values that they do. Yet our existence does not just depend on all of these constants having the right values one by one. It depends on all of them being exactly right in combination. One cosmologist, Lee Smolin, estimates that the probability of all these constants of nature together having values that allow for the emergence of our kind of life is as low as one in 10^{220}, a number that is greater than the total number of atoms in the universe.[27]

How do scientists look upon this fine-tuning of the universe for the emergence of life? Well, some believe that the universe is a product of blind laws of nature that link the different constants in a way that we have yet to understand. Others however embrace some form of what is called the anthropic principle to explain this fine-tuning. The weak form of this principle says that the universe must be life-friendly, since it happens to be inhabited by living observers like ourselves. In the strong form, on the other hand, which is embraced by only a minority of physicists, the anthropic principle states that the emergence of life is inevitable given the values of the basic constants of nature.[28] What I will advocate here is yet another form, which with such standards would have to be referred to as the superstrong anthropic principle: *The purpose of the universe is to generate life, and its physical nature is entirely subordinated to this purpose.* I will however call this the common sense anthropic principle, since it is hard to see that there would be something extreme about postulating that life is not an accident. Such basic principles of science play important roles for how cosmological investigations turn out, and naturally the output of these reflect their input. In science the intention of an investigator also plays a significant role in what questions are asked and what answers are found, and a basic principle of science may allow for solving a problem simply by providing a new context.

THE THEORY OF EVERYTHING AND THE ORGANIZATIONAL HIERARCHY OF LIFE

Ever since Isaac Newton formulated his theory of gravitation in *Principia* in 1687, physics has been perceived as the most fundamental of the sciences and has set a standard that the others have sought to emulate. Newton's theory has in fact proven to be very exact in predicting the mechanics within the solar system, and with regard to the cosmological principle, it has thus been assumed to be valid always and everywhere. Based on this success, physicists have worked in the last three centuries to develop field theories for other forces that would be valid always and everywhere throughout the universe, and this work has provided our understanding for, among other things, the four forces of nature.

It has turned out however to be an insurmountable challenge to develop unified field theories that are applicable on several different levels of the universe, which would amount to the fulfillment of the dream of the realization of the theory of everything that many physicists aspire to. Hence while Newtonian mechanics seems to provide a near-perfect mathematical description of the movements of matter at the level of the solar system, it does not seem to do so at the level of elementary particles. The behavior of elementary particles is instead best described by quantum mechanics, which it has not been possible to unify with relativity theory. To complicate the unification of the different theories further, the existence of very high proportions of "dark matter" and "dark energy" has had to be inferred to explain the observed movements of matter on the galactic and universal levels. Without very significant modifications, Newtonian mechanics is not applicable either on the subatomic level or on the galactic and is only directly applicable on the level where it was first discovered, namely that of the solar system.

Despite the fact that many physicists for some time have claimed that a unified theory of everything is just around the corner, the formulation of one has not yet happened, and I think there are reasons to ask if there could be something fundamentally flawed with the basic assumptions under which science is operating. In the new big bang theory proposed above, it was suggested that the initial rate of expansion of the universe and its early inflationary phase were caused by the emergence of the Cosmic Tree of Life with its axes in three directions. On my own part, I thus believe that it is only with such a fundamental space-time organizer of the universe that is Platonic in nature can the enigmatic fine-tuning of the constants of nature be explained. The Cosmic Tree of Life provides the context for the universe to generate life. In such a context it would make perfect sense that the constant expressing the initial rate of expansion of the universe has exactly the value that it does, simply because this is the one that is conducive to life. From such a context it would also be necessary for the forces of nature to have the particular strengths that they do, and so within such a context it would well be possible to formulate a theory of everything. These relationships we will look into in more detail in chapter 5. In established science, no such context is postulated for the various laws, constants, and forces of nature and for this reason it cannot explain the coherence of the universe. Instead, the laws of nature are thought of as if they had come into existence more or less by accident, and so it is no wonder that a science based on such assumptions has failed to formulate a unified theory of everything. To be able to formulate such a unified theory, it is necessary to have a view of the universe in which laws are expressions of a unified, purposeful context, and if the existence of such a context is denied, then it is only logical that it cannot be formulated.

If we accept the commonsense anthropic principle—that the purpose of the universe is to generate life—then it may also be argued that a necessary approach to a complete theory is one of systems theory. What I am suggesting here is that in order for life to emerge in a universe, the universe needs an organized hierarchy of different levels that have distinct and separate roles in the life-generating process. To fulfill these roles, each of these organizational levels needs to retain a certain level of autonomy. If we accept that the physical laws of the universe are subordinated to its purpose of generating life, maybe laws equally applicable at all its different levels in fact could not exist, and that different laws by necessity would have to apply to elementary particles and galaxies. Moreover, structures at each of these organizational levels need to be defined by Halos and dominated by different mechanical laws to keep the universe from falling back into a homogeneous disorganized state, which would preclude the emergence of life. If we accept that the overall context for the evolution of the universe is provided by the Cosmic Tree of Life, then it is not so difficult to accept that the contexts for lower-level systems in the universe are similarly provided by lower-level Trees of Life with their associated Halos.

The primary systems, or organizational levels, generating the context for life in the Halographic universe are (1) the universe, (2) the galaxy, (3) the solar system, (4) the planet, (5) the biological organism, (6) the cell, and (7) the atom. A well-known example of one such life-organizing system is planet Earth as described in the Gaia hypothesis, a system in which homeostatic mechanisms, seemingly controlled by feedback loops, ensure the stability of the conditions for life. All of these organizational levels rotate around Trees of Life (except the organismic), which are generated by Halos that create open spherical systems (figure 1.5). All of them also have conspicuous, very dense centers, such as the black hole in the galactic center, the cores of the Sun and the Earth, and the nucleus of the atom. Presumably there is also such a core in the center of the universe. These organizations of life on different levels are also, as we shall see in chapter 6, entangled. This entanglement between different Halos ultimately goes back to the birth of the different levels of Trees of Life in the inflationary phase that followed upon the big bang. The Cosmic Tree of Life in three dimensions notably is directly reflected also at the lowest atomic level through the perpendicular electron clouds called orbitals.

I think that it is possible to make the case that without such a hierarchical organization, where each level operates according to rules of its own and has a certain autonomy, life would not have emerged anywhere in the universe. In fact, the relationships between these different hierarchically organized Halos seem to have been fine-tuned, from the universe in its entirety down to the elementary particles, in order for life to emerge. The generation of life thus mandates that the universe be heterogeneous from the very beginning and operate according to different sets of rules on different hierarchical levels, which allows them to maintain their autonomy. If, for instance, stars were not organized in galaxies, the higher elements, such as oxygen, carbon, and metals, would not have become available for the formation of planetary systems in which life could evolve. If planets like our own Earth were not organized into star systems, they would not have been provided with a sustained source of light and heat that is necessary for the emergence of life. If biological organisms did not live within planetary ecosystems, where the basic elements and nutrients are circulated through the oceans and atmosphere, they would not survive. The emergence of life in the universe critically depends on the existence of a hierarchy of organizational levels where the material exchanged between them is narrowly defined.

The reason that Newtonian mechanics, or any form of mechanics for that matter, is not applicable both at the microscopic quantum

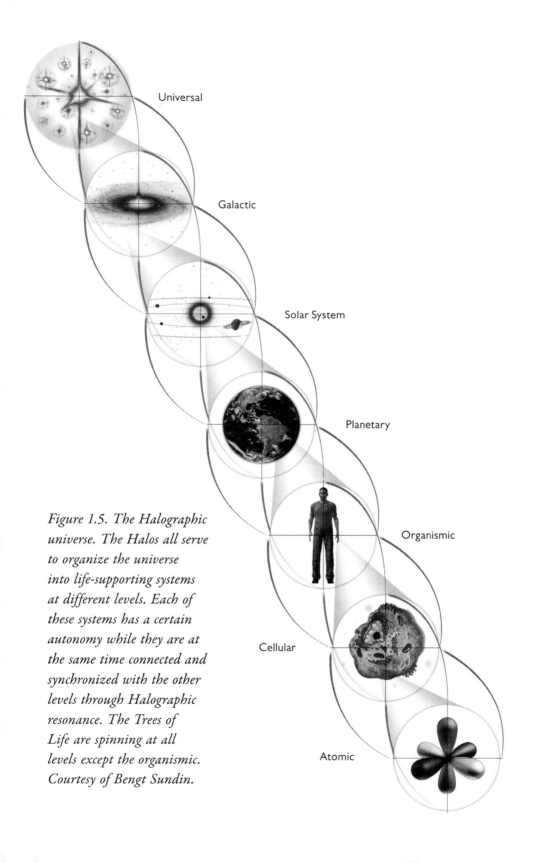

Universal

Galactic

Solar System

Planetary

Organismic

Figure 1.5. The Halographic universe. The Halos all serve to organize the universe into life-supporting systems at different levels. Each of these systems has a certain autonomy while they are at the same time connected and synchronized with the other levels through Halographic resonance. The Trees of Life are spinning at all levels except the organismic. Courtesy of Bengt Sundin.

Cellular

Atomic

level and the macroscopic galactic level (at least not without very serious qualifications such as the assumption of "dark matter") may be that this would not be consistent with the commonsense anthropic principle. For life to emerge, matter needs to be organized independently at each level, and the organization at different levels need to be coordinated and operate according to different mechanical laws. If, for example, the atomic world was not formed by electron clouds in accordance with the rules of quantum theory (in this case orbital theory), a range of distinct chemical elements would not be formed and there would be no life. If, on the other hand, our own level of existence was dominated by such electron clouds, it would be very intangible. What I am suggesting is that the nested hierarchy of the universe allows for the emergence of autonomous life forms because different physical laws dominate the different levels. After all, to have a relative autonomy versus the environment is a fundamental characteristic of a living organism, and maybe the whole universe requires similar autonomy.

It is quite possible that many physicists would disagree with me about this, based on their expert knowledge of their respective specialized fields. The point to realize however is how profound the influence of the basic assumptions of science is on the direction it takes. The researcher who believes that physical laws principally determine the evolution of the universe, and that the emergence of life is an accidental by-product of these laws, will naturally seek to unify the laws of the universe without consideration to the nested hierarchy of systems that is necessary for the generation of life. A researcher like myself, however, who sees the creation of life as the primary purpose of the universe and regards all the physical laws as subordinated to this end, will, on the other hand, regard each of its levels as to some extent autonomous and study how the particular mechanical laws *that apply on each level* have served the emergence of life. The two different approaches will lead the researcher in two very different directions, and it is not possible to go in both directions at the same time. A choice of approach has to be made. What is proposed here is that the existence of a nested hierarchy of systems, as shown in figure 1.5, is a necessary prerequisite for a universe to be able to generate life. The *relationships* between these systems need to be taken into account in any true theory of everything.

Even though there is a relative autonomy of the different levels of organization of life, these are in the theory presented here also entangled and in resonance with each other. The alignment of galactic spins with the Central Axis is one example. The Halographic resonance between biological organisms, planets, and galaxies is another, and this will be discussed in chapter 6. The Halographic resonance of human beings with the planet has been extensively described,[29] and there is no doubt as to the reality of this phenomenon. Such instances of Halographic resonance are among the most fascinating aspects of how the universe functions. Such resonance is behind all spiritual guidance, including premonitions and dreams. I am convinced that if we are to understand how the universe is created and evolves, science needs to focus on the study of how the different structures at different levels of the universe are entangled and how this entanglement serves the generation of life.

In this chapter we have seen how the Cosmic Tree of Life is the fundamental organizer of three-dimensional space in this universe. That indeed such an organizing principle exists I find infinitely more intuitively attractive than the current paradigm of science, which proposes that the evolution of the universe and its biological organisms is a result of a number of natural laws that exist without any context. Here we have now discovered the spatial context provided by the Tree of Life. Evolution is however also fundamentally related to time, and when studying evolution it is equally important to understand how the Cosmic Tree of Life organizes time. Generally speaking, time is to modern people a more mysterious phenomenon than space, and to understand it we will need some guidance from the past.

2

Time and the Calendars of the Maya

THE TWO ASPECTS OF TIME

If we look at the various constants that, according to modern physics, broadly define the nature of our universe, such as the six numbers given by Martin Rees, we may notice that something is conspicuously absent among them: They do not include any constant of time. Even though it is widely recognized that the universe has evolved, or at least changed, over time, demonstrating certain patterns from the big bang to the present time, no constant expressing the rate of this evolutionary process has been included as a primary characteristic of our universe. The absence of a constant of time, or a rate of change, in fact probably reflects how confused we modern humans are when it comes to understanding the nature of time.

To understand time we may first consider that the ancient Greeks recognized two aspects of time, Chronos and Kairos, where Chronos meant "quantitative, measurable time" and Kairos was the "right time," qualitatively speaking.* Chronos could be measured by comparisons with mechanical devices and astronomical cycles, whereas Kairos was perceived only at certain critical instances as an opportunity or a crisis, as if some evolutionary process was trying to force itself into existence at that very point in time. This was then considered as the "right" time. Kairos thus expresses when something is "in the air" and about to happen, and without an understanding of its relationship to evolution it is usually experienced as a magical, or at least subjective, aspect of time. Kairos, or "evolutionary time," is however the aspect of time that gives us the actual *experience* of time as something more than just a series of discrete, unrelated moments. Most of us recognize experiences related to Kairos, for instance in so-called synchronicities, a term coined by the Swiss psychologist Carl Jung to describe remarkable coincidences that statistically speaking would be extremely unlikely.

Officially, the modern world recognizes only the Chronos aspect of time. Many are in fact infatuated with this quantitative time and with always doing things as quickly as possible. We are obsessed with time in the sense of this measurable mechanical time, and we honor athletes who are the fastest in their respective sports and tend to favor cars that can accelerate to a certain speed in the shortest time possible. We appreciate computers that can accomplish a task in the "shortest" possible time because this "saves" time, and yet, because of this emphasis on Chronos, many people today are stressed out because they do not have "enough" time. (Often those who are the most into "saving" time are in fact also those who have the "least" time at their disposal.) Mechanical cycles are used to make quantitative determinations of what

---

*Sometimes you hear linear and cyclical time being contraposed. It seems to me however that linear and cyclical are really one and the same mechanical, quantitative aspect of time. Thus as soon as you have completed one astronomical cycle, such as a year, you may start to count such cycles, year 1, year 2, year 3, and so on, and it turns into linear time. Cyclical time and linear time is then really the same thing. Evolutionary time, or Kairos, is different from this.

is "too little," "enough," or "too much" time, and ultimately mechanical time is measured against some astronomical cycle in our local solar system, such as the revolution of the Earth around the Sun (the year) or the Moon around the Earth (the month). Modern physics has become very adept at working with the Chronos aspect of time, and its standard unit, the second, is today defined by atomic vibrations.

Yet since the modern world recognizes only the measurable aspect of time, whenever we are part of remarkable coincidences, we tend to regard them as mysterious. In fact, such remarkable coincidences often turn out to be the defining moments of life that serve to guide its future direction. Indeed, the reason many place so much weight on such events is, as we shall see, that they stem from a higher evolutionary aspect of time, Kairos, instead of the common measurable aspect. The two separate aspects of time also account for why time often seems like an area where our measurements and experience diverge (*time flies when you are having fun*). The divergence stems from the fact that most people would probably deny the Kairos aspect of time or look upon it as something merely subjective that has its origin in the human psyche. With society's emphasis on the Chronos aspect of time, many look upon synchronicities (*"she called just when I was thinking of her"*) as something out of the ordinary whose existence they do not understand. In general, we look upon time as something clearly more mysterious than the three dimensions of space, and many aspects of time and timing have become blind spots in our modern civilization mainly because we have denied the very existence of Kairos. Kairos expresses the "timing" of the cosmos, and obviously whenever we are part of this we have reasons to acknowledge that we are part of a higher purpose.

The denial of Kairos, the evolutionary timing of the cosmos, helps explain why modern physics has not included a constant of time in its most basic description of the universe. Mechanical cycles upon which quantitative time is based cannot serve as true constants of nature since they change over time. The year and the month, for instance, upon which most calendars are based, have had a meaning only for the past five billion years or so. Before this time the solar system had not even come into existence, and its cycles have varied considerably in duration since it did. Yet the evolution of the universe has happened "in time" ever since the big bang, and we may ask if there is not any way that we could quantitatively express the speed of this process. Even if Kairos has always been perceived as something subjective, or even magical, there is a way of looking at this aspect of time as an expression of the overall evolutionary rate of the universe at any given time. The "right time" would then be when such an evolutionary opportunity forces itself into existence even if this may be experienced very subjectively. What is suggested here is that the reason that such evolutionary opportunities turn up to begin with is that some objective, yet invisible, factor drives these opportunities to manifest themselves at the "right time." This factor would then have to be monitored as evolutionary "right time" rather than as mechanical time. As examples of such "cosmic timing" of evolutionary opportunities, we may take the relatively frequent independent and simultaneous discoveries in science, as well as technology, like those of Leibniz and Newton of calculus or Bell and Gray of the telephone.* If we do not want to look upon such synchronicities as mere curiosities, we have to conclude that some factor exists that serves to synchronize events in the universe and has quite some defining power over our lives.

Another good example of this synchronizing power of Kairos is the big bang, the primordial synchronizing event, which all other events in evolutionary history go back to, when in less than a minute the four basic forces of nature—energy, matter, three-dimensional space, and time—all came into existence. Few physicists have stopped to consider the amazing fact that all of these phenomena actually emerged in *synchrony* in the big bang. If the universe had no purpose or intelligence, these different things might just as well have appeared at random moments

*Alexander Graham Bell paid his application fee for the telephone patent only a few hours before Elisha Gray on February 14, 1876.

spread out over a long period of time to create an incoherent universe, but instead their emergence was synchronized and coordinated. All evolutionary time goes back to this synchronized initial event, but the mechanical cycles, and the Chronos measurable time based on them, are in fact secondary phenomena coming into existence much later. Astronomy, biology, and geology are, as we shall see, examples of how different phenomena emerged at the same time in a synchronized way to provide the conditions for life, and this would not have been possible if all evolution did not go back to the same starting point.

To accept such a synchronizing factor behind evolution is not necessarily easy for modern people, who mostly only follow mechanical time and regard everything else as anomalies. Yet it does seem clear that the Greeks saw it differently. A people who saw time *even more* differently from the modern world were the ancient Maya, who not only recognized the existence of synchronicities in many different ways on their steles, but actually even developed a calendar system that was designed to chart Kairos, or evolutionary time, as this emanated from the Tree of Life. It is because they charted this "other" aspect of time that I believe it is imperative to make the Mayan calendar system part of any attempt to develop a new and truer theory of the evolution of the universe. In the Mayan view time had its origin in the Tree of Life, whose time was what their calendar system was designed to chart. By looking at the Mayan calendar system, we will come to understand that the universal process of evolution was initiated through the synchronizing power of the Tree of Life at the big bang, but also that this synchronizing power would be manifested at many later points as well to maintain the direction of evolution. From this we will be able to conclude that the Tree of Life is not only the organizer of space, but also of time, and the empirical evidence for this will be provided as we go along. If the creative pulses of the Tree of Life are correlated with mechanical time, this Kairos aspect of time emanating from the Cosmic Tree of Life can be followed by calendars, and this correlation is what the Mayan calendar priests would devote great efforts to bring about.

Even though the ancient Greeks, unlike modern people, recognized both aspects of time, they never developed a calendar system with which the two could be followed in parallel. To have done so is the unique contribution of the Mayan civilization to our planet, and there are reasons to ponder what modern science would look like today had it found its roots among the Maya, who went more deeply into the matter of time, rather than the Greek. As the reader will probably gradually find, the answer is that it would be very different. The science, which will emerge in this book, does indeed owe some significant lines of thinking to the Maya, with which we will be better equipped to understand the origin of the universe and of ourselves.

THE MAYA AND THEIR RELATIONSHIP TO TIME

So how are the different evolutionary processes of the universe synchronized, and how can we follow these processes in time? Well, as mentioned above, they are not directly measurable as astronomical or mechanical cycles, and so the common Gregorian, Muslim, or Jewish calendars that are based on such are of no help to us in understanding Kairos. Even if we see glimpses of evolutionary time in other ancient civilizations,* the Maya are the only people of this planet who have systematically charted it and correlated it with mechanical time. It is in fact often said that the Maya were obsessed by time, and one thing that sets this people apart from the other civilizations of our planet is that they lived in a system of rule that may be referred to as "calendrocratic." In the excellent *Maya Political Science*, Prudence Rice describes how the Mayan civilization strove for their political system to be organized in harmony with the calendar. Life, in other words, revolved around, and was ruled by, the calendar to an extent that people who live today may find very difficult to grasp. This calendrical fervor is however also the

---

*In ancient times China had a religious oracle bones calendar based on a 360-day period as well as a civil calendar of 365 ¼ days.

very reason that Mayan cosmology will today by necessity play a critical role in any serious attempt to understand the evolution of the universe.

According to Rice, even the name Maya is derived from a calendrical word, *May,* meaning "cycle of thirteen katuns," so that Maya really means "the people of the cycle of the thirteen katuns" or alternatively "the people of the four directions."[1] The latter is a reference to the World Tree, whose axes to the Maya are the source of the sacred directions both on Earth and in the cosmos. The naming of this people as Maya then tells us that even their very identity to a high degree was based on the calendar, and to royal dynasties the calendar also served as a significant socially unifying factor. The political system of the Maya was directly organized by the calendar in that it defined a system for rotating power between cities at some of its significant shift points. The calendar described the different qualities of time generated by the World Tree, and so if there had been no connection to the World Tree, the prophetic calendar would not be seen as serving a purpose. Calendrical shift points were extensively celebrated, and at such occasions prophets, officially designated for specific time periods, were "seated" and would give guidance to the cities based on their calendrical knowledge. The political power rotated between the different capitals of the north, west, south, and east in order for the system of rule to be in harmony with the various directions created by the World Tree. Thus in this culture there was an inseparable relationship between the World Tree and the calendar.

The various days and time periods of this calendar may be described as a code made up essentially of numbers, day-signs, and deities, with different qualities symbolic of the respective time periods. Literally everything in the Mayan culture was seen through the filter of the calendar and understood with respect to the various deities and day lords that were believed to create the spiritual qualities of time. In all their significant doings they would consult their Sacred Calendar, and many of their descendants still do. Their diet as well as the healing of different diseases was based on the calendar. A child was typically given its name from the calendrical day when it was born, and it seems from ancient Mexican codices[2] that people would relate to one another to a large extent based on the identities that were associated with these calendrical energies. A distinct class of astronomer-priests and scribes were deeply engaged in studying evolutionary time cycles and correlating the celestial movements to this calendar. While they took a lot of interest in observing the astronomical cycles, those were considered secondary to the nonastronomical, such as the 260-day Sacred Calendar.

Unlike in our modern culture, the Maya did not keep their science separate from the spiritual aspects of their existence, and they rather saw science as a means of understanding divine creation. Their science was about understanding how space (the four directions) and time (the calendar) emanated from the World Tree and how prophecy could be based on this knowledge. It is because of this that their calendrical cosmology is important today to those who want to capture the "other" aspect of time and understand evolution.

A BRIEF HISTORY OF THE MAYA

Who then were—and are!—the Mayan people? Before describing the workings of their calendar system I feel it is important to present a brief background history of this people, who have often been described as mysterious and enigmatic. For those desiring to further study the culture of the Maya, I would, in addition to the book by Rice, like to recommend the work by Mayanists Linda Schele and David Freidel, which focuses on the dynastic life[3] and the spiritual life[4] of the Maya. My own earlier books also discuss the calendar system at length.[5]

The Maya were, and are, a Native American people who for some five thousand years cultivated maize in what is now Guatemala, Honduras, Belize, and southern Mexico, and who emerged with an advanced culture approximately in synchrony at the time of the foundation of the Roman Empire and the teachings of Christ. In the third century CE their cities and temple sites mushroomed in Central America, and these city-states had a number of typical characteristics. They were for instance ruled by separate dynastic lineages of *ahauob* in each of the different major cities. We might today refer to these rulers as shaman-kings (the word *ahau*, as we noted earlier, means "lord"). They were regarded as personifications of the cosmos and were responsible for performing the many calendar-based ceremonies that were a significant part of life among them. The Maya also had scribes and were in fact the only people in pre-Columbian America to have a written language. Following the decipherment of the Mayan writing system by the Russian scholar Yuri Knorozov in the late 1940s[6] it has been possible to read much of the stone inscriptions at the pyramid sites, and a fairly good picture of life among the ancient Maya has since been gained.

During the period lasting from the third century to the beginning of the ninth century CE, the Mayan civilization was one of the most significant in the world. Toward the end of the seventh century, at a time when Paris and London were mere villages, the city of Tikal had around eighty thousand inhabitants and impressive pyramids, ball courts, and altars, while its whole city-state harbored five hundred thousand people.[7] The whole Mayan region is estimated to have had about eight million inhabitants around this time and was engaged in trade with faraway areas, especially with the significant city of Teotihuacán in Central Mexico.

In the early ninth century, a significant shift took place in the Mayan calendar (the beginning of the eleventh *baktun* in the Long Count in CE 830), and at this time the Classical Mayan culture in the south collapsed. Traditional historical and archeological science has sought the reason for this "disappearance of the Maya" in material reasons ranging from internal warfare or invasions to the currently more popular climate change.[8] My own view of it, as described in detail in previous books, is that it was this very calendrical shift point in evolutionary time that ultimately caused the collapse of the Classical Maya, even though fighting between city-states as to which one would be the capital of the new baktun may have exacerbated the demise. Regardless

Figure 2.1. The Dresden Codex. The Dresden Codex is considered to be the finest of the four currently known Mayan codices from pre-Hispanic times, which are all calendars. The page shown is from the Venus tables, which describe how the cycle of Venus is related to the Mayan Sacred Calendar. Based on studies of this codex, the German librarian Ernst Förstemann was able to elucidate much of the Mayan calendar system.

of how it happened, at that time almost all the Classical sites, such as Palenque, Copán, Tikal, Yaxchilan, Calakmul, and Quirigua, were relatively abruptly abandoned. The center of gravitation of Mayan culture instead moved north to the Yucatán Peninsula, where Chichén Itzá became a new capital. This is today the most visited Mayan site and was recently voted[9] one of the New Seven Wonders of the World. As the Mayan civilization then transited to its so-called Postclassical phase, the system of rule by competing dynasties of ahauob came to an end and so did the use of its more advanced long-term calendars. As Chichén Itzá was deserted in CE 1224, the Postclassical phase came to an end, and in the era that followed, the Mayan civilization seems to have become less grandiose, based more on local townships in which spectacular pyramids or ball courts were no longer built.

Around CE 1480 a man by the name of Chilam Balam (meaning "Jaguar Prophet") appeared, who predicted the imminent arrival of a new civilization from the east that would establish a new religion. These prophecies were pronounced shortly before Columbus (from a European perspective) discovered America in 1492 and Hernán Cortés, the Spanish conquistador, landed in Veracruz in 1519 and subsequently conquered the Aztec Empire in Central Mexico. In 1526 Cortés launched his Honduran expedition, which defeated the Maya, forced them to convert to Christianity, and imposed a harshly regimented rule on them.[10] In 1562, most of the books of the Maya, which presumably were all calendars (see figure 2.1), were burned at the order of Catholic Bishop Diego de Landa as the "work of the devil," which apparently to his surprise caused much grief to the natives.[11] Their World Tree then came to be transformed into the Christian cross, and its original meaning was essentially lost.[12] Still to this day, however, descendants of the ancient Maya in Guatemala hang branches from the holy Ceiba tree on crosses at Christian burial sites as a reminder of the original symbolic meaning of the cross as the World Tree.

The fate of the Maya is similar to that of so many other Native American peoples in the confrontation with the guns, steel, and germs of European civilization. As a result, the population of Mexico, according to some sources, declined from twenty-five million to one million over the course of a century.[13] This was compounded by a cultural genocide in which the calendars of the Maya initially were branded as "idolatry" and later dismissed by archaeologists and tour guides as superstitions. The plight of the Maya continued under the colonial rule of the Spanish and after independence under different military regimes. It may in fact be only in the year 2008 that a president has been elected in Guatemala who gives recognition to its indigenous population.[14]

As a result of these political and religious developments, the use of the Mayan calendar was forced underground about five hundred years ago, and most aspects of it went out of use. Yet its core, the 260-day Sacred Calendar with its day-signs and numbers, was kept alive by daykeepers over the centuries that followed. These daykeepers were responsible for preserving the ancient calendrical knowledge and for forwarding it to new generations of adepts. As anthropologists visited the mountain villages of Guatemala in the 1960s, they discovered that the daykeepers in these areas still used the same ancient day-signs as were found at the pyramid sites of the Classical Maya. To their amazement they also found that if they calculated back from the calendar the daykeepers were still using to the symbols on the inscriptions at the ancient pyramid sites, not a single day had been lost. This meant that the Sacred Calendar had been passed down without interruption for a very long time. The daykeepers had been imbued with the value of their calendar from their mentors, who had conveyed a sense that its knowledge was important for future generations and sometimes even more broadly for all of humanity.

Additional information about the calendar system has been gathered from the four surviving codices (books) from Postclassical times. The *Book of Chilam Balam*,[15] written after the Conquest, also tell us much about how the Maya viewed their prophetic calendar system based on *katuns, tuns,* and *kins.* It was however only in the 1970s that the first pioneers in the modern world, such as Shearer,[16] Waters,[17] and Balin,[18] began to realize that the Mayan calendar held a profound truth that the rest of the world had been oblivious to, and that this could only be understood by acknowledging an entirely different concept of time. Since then, increasing numbers of people have become aware that the Mayan calendar system holds information crucial to the understanding of the universe we live in, information that can hardly be gained from any other source. What fundamentally makes the Mayan calendar different from any other is that it captures the "other aspect of time," evolutionary time, which reflects the large-scale change, or the "timing" of the universe. Their prophetic calendar is not astronomically based, and this is why it has such a special role to play. We will now look at how this calendar is organized.

THE PYRAMIDAL STRUCTURE OF THE NINE UNDERWORLDS

Most Mayan temple sites have in a central location a pyramid built in nine stories. Such pyramids have been erected in Chichén Itzá, Palenque, Tikal (figure 2.2), and Uxmal, and they played important roles in the celebrations of calendrical shifts, such as the beginning of new baktun or katun periods. These pyramids speak of the predominant role the number 9 had in the cosmology of the ancient Maya. In the Mayan calendar system 9 is the number of Underworlds. Each of these Underworlds was "ruled" by one of the *Bolon-ti-ku,* the nine "gods" of time, and to understand this we must translate the concept of an "Underworld" into a "Level of Evolution" or "Level of Creation," since the Maya held that there were "several creations" as part of the whole. The only Mayan stele that talks about the end of their calendar, monument 6 in Tortuguero, describes this as the "descent" of nine creation gods. What this means is that at that point in time the nine Underworlds will be fully manifest and their corresponding "gods" (in modern language really "cosmic forces") will manifest. A nine-storied pyramid is hence a symbol of the large-scale cosmic evolution scheme in its entirety. To climb such a pyra-

Figure 2.2. The Pyramid of the Jaguar in Tikal. This pyramid was inaugurated in CE 692 and is the highest pyramid in the Americas, reaching to a height of 44 meters. Such pyramids built in nine terraces are typically located in the centers of Mayan temple sites and were used for celebrating shifts between different katuns. (Photograph by the author.)

mid, as the ancient shaman-kings would do, was a means for them to symbolically elevate themselves to the highest level of this scheme. In this scheme the evolution of each Underworld is built on its underlying levels, and the nine Underworlds are not cycles that follow upon, or replace, the preceding one. Instead, each level, or Underworld, not only builds on the lower levels, but also provides the foundation for the higher levels of evolution (figure 2.3).

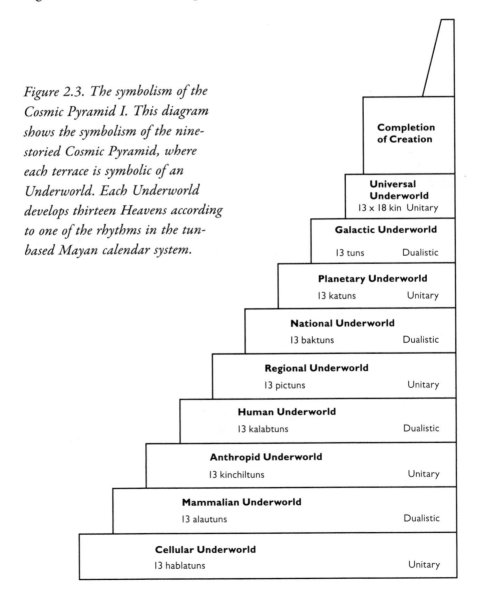

Figure 2.3. The symbolism of the Cosmic Pyramid I. This diagram shows the symbolism of the nine-storied Cosmic Pyramid, where each terrace is symbolic of an Underworld. Each Underworld develops thirteen Heavens according to one of the rhythms in the tun-based Mayan calendar system.

Completion of Creation

Universal Underworld
13 x 18 kin   Unitary

Galactic Underworld
13 tuns   Dualistic

Planetary Underworld
13 katuns   Unitary

National Underworld
13 baktuns   Dualistic

Regional Underworld
13 pictuns   Unitary

Human Underworld
13 kalabtuns   Dualistic

Anthropid Underworld
13 kinchiltuns   Unitary

Mammalian Underworld
13 alautuns   Dualistic

Cellular Underworld
13 hablatuns   Unitary

Each of these nine Underworlds, or levels of evolution, is in turn subdivided into thirteen Heavens, thirteen qualitatively different periods of time that are activated in a special sequential order. Both Underworlds and Heavens, and their corresponding "gods," really refer to cosmic forces dominating specific periods of time. This, of course may be confusing to us, as we would spontaneously think of Heavens and Underworlds as spatial concepts and gods as individuals. The durations of these so-called Heavens were the same in a given Underworld, but different in different Underworlds (table 2.1), which then in turn came to dominate for different lengths of time. In this table the time periods that these Heavens dominated in the different Underworlds are shown together with their Mayan names, which are also translated to mechanical time (Chronos). Each of the nine Underworlds develops according to a constant rhythm of its own, which is based on the frequency of the most significant evolutionary shifts between the Heavens of that Underworld. The lowest Underworld, here going by the name of the Cellular Underworld,

TABLE 2.1. IMPORTANT TIME PERIODS OF THE MAYAN UNDERWORLDS

Underworld	Mayan Time Period	Kairos Time	Chronos Time
9th	oxlahunkin	18 kin	18 days
8th	tun	$20^0 = 1$ tun	360 days
7th	katun	$20^1 = 20$ tun	19.7 years
6th	baktun	$20^2 = 400$ tun	394.2 years
5th	pictun	$20^3 = 8,000$ tun	7,900 years
4th	kalabtun	$20^4 = 160,000$ tun	158,000 years
3rd	kinchiltun	$20^5 = 3,200,000$ tun	3.2 Myr*
2nd	alautun	$20^6 = 64,000,000$ tun	63.1 Myr
1st	hablatun	$20^7 = 1,280,000,000$ tun	1.26 Gyr*
*Myr = millions of years, Gyr = billions of years			

is subdivided into thirteen Heavens that each last for a *hablatun* of exactly 20⁷ *tun* (a tun is a period of 360 days), or approximately 1.26 billion years (figure 2.3). This Underworld then spans 13 × 1.26 = 16.4 Gyr (billion years) in its entirety. As another example, each of the thirteen Heavens in the National Underworld last for a baktun, 20² tun exactly, or about 394 years, which then in its entirety will span 13 × 394.2 = 5,125 years. A date described in terms of these different levels, each of thirteen Heavens, is shown at Stele 1 in Coba at the Yucatán Peninsula (figure 2.4). As the pyramid is climbed, the frequencies increase twenty times every time a new higher level, or Underworld, is activated (except at the highest level of evolution).

To begin our study of the reality underlying this calendar system, we may look at events in cosmic history that according to modern datings (see table 2.2 on page 47) occur as each of these nine Underworlds were initiated. What then becomes clear is that very significant events in evolution indeed took place close to the activation times of the Underworlds; the first emergence of matter in the big bang (Underworld 1), the first human beings (Underworld 4), and the first advanced civilizations (Underworld 6), to name only a few. In general terms the four lowest Underworlds are initiated by events in material and biological evolution, the next four Underworlds by events in historical and mental evolution, while the nature of the ninth one, to begin March 8, 2011, is yet unknown. Table 2.2 also shows the general nature of the phenomena that each Underworld develops.

According to Mayan myth their calendar is charting pulses originating from the Tree of Life, the central creative principle of the universe. The rhythms with which these Underworlds develop are not based on astronomical, or mechanical, cycles, such as months, years, or precessional rounds, but instead define wave movements of evolutionary time. Although the basic time period, the tun of 360 days, is close to the solar year of 365.2422 days, the use of the tun as the base in the prophetic calendar system was not the result of a mistaken estimate of the duration of the year on the part of the Maya. The Maya made the most

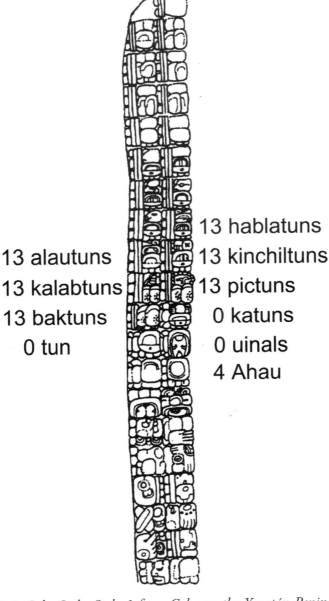

13 alautuns
13 kalabtuns
13 baktuns
0 tun

13 hablatuns
13 kinchiltuns
13 pictuns
0 katuns
0 uinals
4 Ahau

Figure 2.4. Coba Stele. Stela 1 from Coba on the Yucatán Peninsula. Coba is one of the few Mayan sites that were inhabited both during the Classical and Postclassical times. This stela gives the creation date in terms of several different series of thirteen time periods in the tun-based system. Courtesy of Linda Schele and David Freidel.

TABLE 2.2. DURATIONS, INITIATING EVENTS, AND CHIEF PHENOMENA OF EACH OF THE UNDERWORLDS

Underworld (Duration)	Initiating Date	Initiating Event	Phenomena Developed
Universal (13 oxlahunkins)	March 8, 2011	Oneness field	?
Galactic (13 tuns)	Jan. 5, 1999	IT civilization	Information technology, galaxy
Planetary (13 katuns)	July 24, CE 1755	Industrialism	Industrialism, materialism, democracy, republics, electricity, planet
National (13 baktuns)	June 17, 3115 BCE	Higher civilizations	Written language, major constructions, historical religions, science, fine art, monarchies, nation
Regional (13 pictuns)	103,000 YA*	Spoken language (?)	Spoken language, complex tools, agriculture, shamanistic religion, cave art, chiefdom
Human (13 kalabtuns)	2.05 MYA	First humans	Various forms of *Homo*, tribes
Anthropoid (13 kinchiltuns)	41 MYA	First monkeys	Monkeys, australopithecines, family
Mammalian (13 alautuns)	820 MYA	First animals	Animals, plants, multicellular organisms
Cellular (13 hablatuns)	16.4 BYA	Big bang, first matter	Galaxies, stars and planets, chemical elements, unicellular organisms

*YA = years ago; MYA = millions of years ago; BYA = billions of years ago

accurate estimate of the solar year known from their own time (not surpassed until the scientific revolution in Europe in the seventeenth century) of 365.2420 days, and in their agricultural *Haab* calendar, which had another purpose, they used 365 days. Thus, they knew that the tun was different from the solar year, and so the use of the tun as the base unit for their prophetic calendar system was the result of a conscious choice, and we will gradually understand why.

I would like to add that personally I do not regard the superior knowledge of time of the ancient Maya as a reason to idealize them. As much as other contemporary peoples they were engaged in warfare, and it has been a long time since archeologists upheld the view that they were peaceful stargazers. In fact, among the Maya it seems like warfare, where prisoners were taken to be sacrificed, often took the peculiar form of calendar wars. Since it gave prestige, privileges, and divine endorsement for a temple area to be designated as the capital of a certain direction, or time period (may, baktun, or katun), wars would sometimes break out because of calendrical issues dividing the different city-states. I feel however that this should not deter anyone from learning what this fascinating civilization knew about time.

THIRTEEN HEAVENS

The thirteen Heavens were thus perceived as a special order of spiritual qualities, or heavenly archetypes, that in all of the Underworlds proscribe the sequence of the corresponding thirteen time periods. These Heavens were then associated with a series of thirteen deities— gods and goddesses—that were believed to endow the respective time periods with their special spiritual qualities. These thirteen deities are however not thirteen randomly assigned qualities. Instead, they follow a metaphorical sequence from seed to mature fruit, creating an organic process of evolution within each Underworld, and this will be very relevant as we study biological evolution from a new perspective. Different metaphors for the thirteen Heavens, or at least the seven odd-numbered ones, are given in table 2.3. The Tree of Life may here be looked upon as a set of strings that creates a wave movement with defined frequencies, giving rise to vibrations that use the universe as a resonance board. In this wave movement, the odd-numbered Heavens are peaks of creativity, promoting evolutionary steps forward, whereas the even-numbered Heavens are valleys, periods of rest or integration. While the odd-numbered Heavens usually produce the most evident steps forward in the evolution of any Underworld, the even-numbered

TABLE 2.3. METAPHORICAL DESCRIPTIONS OF THE THIRTEEN HEAVENS

Heaven (Day or Night)	Character of Time Period	Name of Ruling Aztec God or Goddess*
Day 1 (1st Heaven)	Initiation Sowing	Xiuhtecuhtli, god of fire and time
Night 1 (2nd Heaven)		Tlaltecuhtli, god of the Earth
Day 2 (3rd Heaven)	Expansive Push Germination	Chalchiuhtlicue, goddess of water and birth
Night 2 (4th Heaven)	Reaction	Tonatiuh, god of warriors and the Sun
Day 3 (5th Heaven)	Anchoring Sprouting	Tlacolteotl, goddess of love and childbirth
Night 3 (6th Heaven)		Mictlantecuhtli, god of death
Day 4 (7th Heaven)	Midpoint Proliferation	Cinteotl, god of maize and sustenance
Night 4 (8th Heaven)		Tlaloc, god of rain and war
Day 5 (9th Heaven)	Breakthrough Budding	Quetzalcoatl, lord of light
Night 5 (10th Heaven)	Destruction	Tezcatlipoca, lord of darkness
Day 6 (11th Heaven)	Proto-Highest Expression Flowering	Yohualticitl, goddess of birth
Night 6 (12th Heaven)		Tlahuizcalpantecuhtli, god ruling before dawn
Day 7 (13th Heaven)	Highest Expression Fruition	Ometeotl, the supreme deity, lord and lady of duality

*The Aztec gods and goddesses are used because the names of the Mayan deities are unknown.

ones rather generate phenomena that pave the way for these steps. This wave movement may be likened to how daytime activities differ from nighttime activities, and the deities associated with them roughly alternate between light and darkness.

Based on much empirical data, some of which will be presented in this book, I have come to the conclusion that the time periods of the seven odd-numbered and six even-numbered Heavens of the Maya are the same as those referred to as the seven Days and six Nights of God's creation in the Book of Genesis, where God is said to rest on the seventh of these Days. That Genesis does not refer to regular days and nights as is thought by so-called Young Earth Creationists should be immediately clear from the fact that the stars, presumably including our Sun, which is the source of the shifts between regular days and nights, were created in one of these Days.* In the Bible the idea of seven pulses is later reiterated in the Book of Revelation, where the number 7 is mentioned as many as fifty-two times,[19] often as part of what seems like evolutionary progressions. We will therefore refer to the seven odd-numbered Heavens of each major Underworld as Days and to the six even-numbered heavens as Nights, meaning that *each of the Underworlds has its seven Days and its six Nights.*

The Mayan creation story is too long to present in full here for comparison, but the following excerpt from *The Book of Chilam Balam at Chumayel* should give the reader a sense of how it is written: "1 Chuen, The day he rose to be a day-ity and made the sky and earth. 2 Eb, he made the first stairway. It ebbs from Heaven's heart, the heart of water, before there was earth, stone and wood. 3 Ben, the day for making

---

*The separation of light and darkness in the first Day in Genesis corresponds to what we here call a yin/yang polarity. The physical light from the stars and the Sun is a secondary phenomenon that comes later. Genesis 1:3: "And God said, 'Let there be light,' and there was light. God saw that the light was good, and he separated the light from the darkness. God called the light 'day,' and the darkness he called 'night.' And there was evening, and there was morning—the first day." See also Genesis 1:16–1:19: "He also made the stars . . . the fourth day."

everything, all there is, the things of the air, of the sea, of the earth. 4 Ix, he fixed the tilt of the sky and earth. 5 Men, he made everything. 6 Cib, he made the number one candle and there was light in the absence of sun and moon. 7 Caban, honey was conceived when he had not a caban . . ." etc.[20] Like in the Jewish creation story, the course of events has been scrambled compared to how it is known from a modern scientific account of beginnings, but unlike in Genesis the different events in this creation account are linked to calendrical qualities of the Sacred Calendar such as 1 Chuen, 2 Eb, and so on. The idea here is not to say that this provides an accurate description of the course of events, but only to point out that the general approach may be fruitful. The linking of evolutionary processes to calendrical qualities will also be shown to be absolutely necessary, for a modern understanding of evolution.

Nine Underworlds, each made up of thirteen Heavens, which in turn may be subdivided into seven Days and six Nights, provides the basic description of the Mayan calendar system, which in fact is a very simple formula for all evolution. Notably, in this system the most basic Underworld goes back to the time of the big bang, and the Mayan calendar describes a system of nine interconnected divine creation processes. The difference from the ancient Jewish account, which very likely had its origin in Sumer, is primarily that in the Mayan calendar there are *nine* creations, each of seven Days and six Nights, and that these creation processes are not seen as having happened and been completed in the past, but are still ongoing. Although the Mayan calendar system can be studied at a great degree of complexity, it is, initially at least, most useful in this simple form, where it is unified in the simple model of a pyramid symbolizing nine Underworlds, each of which is developed through a progression of thirteen Heavens with the same spiritual qualities. This model of cosmic creation forces can then be used for empirical comparisons with modern scientific accounts. The wave movement of seven Days and six Nights is re-created at different levels according to the following principle: As in the long cycles, so also in the short. This means that the shorter waves are like ripples reflecting the great oceanic waves of the longer Underworlds, or, in other words, that evolution is fractal in nature. In a more complex form the Mayan calendar is really a system of cycles within cycles within cycles.

This structure of the nine Underworlds provides for something we will come back to in chapter 4 as we study biological evolution, namely a periodic system of evolution (table 2.4). In my previous books I have shown such schemes that are applicable to the historic evolution of humankind, and one of them is offered here to illustrate the idea of comparing effects of the different Heavens in the different Underworlds. As a fairly topical example of how to use this periodic system of evolution, we may take Night 5, the tenth Heaven, which in Aztec mythology (the names of the Mayan deities are unknown) was ruled by the Lord of Darkness and usually has a destructive character. In the National Underworld this quality of time correlates with the Dark Ages, when the Roman Empire and its economy were collapsing in the western regions. In the Planetary Underworld the same quality of time coincides with the Great Depression, and in the Galactic Underworld it corresponds with the global financial crisis of 2008, predicted for more than a decade by those who follow the Mayan calendar. (As an illustration of the power of predicting the timing of events offered by the true Mayan calendar, which ends October 28, 2011, in 2003 I concluded the economy section in my book *The Mayan Calendar and the Transformation of Consciousness* with the words: "Regardless of what forms such a [financial] collapse may take it seems that the best bet is for it to occur close to the time that the fifth Night begins, in November 2007 [strictly speaking the 19th]."[21] There is now a consensus among economists that the global recession started in December of 2007, making the prediction based on the Mayan calendar almost perfect.) Hence, the quality of a particular Heaven, or cosmic force, is reflected in all the different Underworlds. The periodic system of evolution is meant to be used for understanding evolutionary time periods, since comparable phenomena tend to occur during the same Heaven throughout the various Underworlds.

TABLE 2.4. THE PERIODIC SYSTEM OF EVOLUTION IN THE FOUR HIGHEST UNDERWORLDS

Ruling Quality	National Underworld of 13 baktuns	Planetary Underworld of 13 katuns	Galactic Underworld of 13 tuns	Universal Underworld of 13 oxlahunkins
1st Heaven is Day 1 Sowing	June 17, 3115–2721 BCE	July 24, 1755–1775	Jan 5, 1999–Dec 30, 1999	March 8–March 25, 2011
2nd Heaven is Night 1	2721–2326	1775–1794	Dec 31, 1999–Dec 24, 2000	March 26–April 12
3rd Heaven is Day 2 Germination	2326–1932	1794–1814	Dec 25, 2000–Dec 19, 2001	April 13–April 30
4th Heaven is Night 2 Reaction	1932–1538	1814–1834	Dec 20, 2001–Dec 14, 2002	May 1–May 18
5th Heaven is Day 3 Sprouting	1538–1144	1834–1854	Dec 15, 2002–Dec 9, 2003	May 19–June 5
6th Heaven is Night 3	1144–749	1854–1873	Dec 10, 2003–Dec 3, 2004	June 6–June 23
7th Heaven is Day 4 Proliferation	749–355	1873–1893	Dec 4, 2004–Nov 28, 2005	June 24–July 11
8th Heaven is Night 4	355–CE 40	1893–1913	Nov 29, 2005–Nov 23, 2006	July 12–July 29
9th Heaven is Day 5 Budding	40–434	1913–1932	Nov 24, 2006–Nov 18, 2007	July 30–Aug 16
10th Heaven is Night 5 Destruction	434–829	1932–1952	Nov 19. 2007–Nov 12, 2008	Aug 17–Sept 3
11th Heaven is Day 6 Flowering	829–1223	1952–1972	Nov 13, 2008–Nov 7, 2009	Sept 4–Sept 21
12th Heaven is Night 6	1223–1617	1972–1992	Nov 8, 2009–Nov 2, 2010	Sept 22–Oct 9
13th Heaven is Day 7 Fruition	1617–Oct 28, 2011	1992–Oct 28, 2011	Nov 3, 2010–Oct 28, 2011	Oct 10–Oct 28, 2011

We may now begin to better appreciate the difference between Chronos and Kairos time. While Chronos time is continuous and can be measured in principle down to an accuracy of many decimals of a second, Kairos time is instead quantized and is an expression of an energy *state* of the Tree of Life with which the entire cosmos is in resonance. Take for instance the Gregorian date November 6, 1632, which means that 12 baktuns, 0 katuns, 15 tuns, 12 uinals, and 3 kin have passed since the beginning of the National Underworld, and so by the Maya this date would in Arabic numbers have been expressed as 12.0.15.12.3. Because each of these numbers corresponds to a specific cosmic force, this is not just a way of marking the passage of time, but this sequence of numbers in a deeper sense reflects the energy *state* of the Tree of Life on that particular day. Mayan time is then fundamentally quantized, and unlike astronomical cycles, which form the basis of Chronos time, it is not continuous. If we are keeping track of the evolution of the universe all the way from the big bang, its current state can be described by a series of quantum numbers, such as 12.12.12.12.12.12.0.15.12.3. The series of quantum numbers in the Mayan calendar describes how many different time periods such as hablatuns, alautuns, and so on have passed since the big bang on the abovementioned day (see table 2.1). While there is a risk in introducing this terminology, since most people associate quantum physics with unusual properties of elementary particles, the terms *quantum* and *quantize* merely refer to the fact that the associated time periods are associated with distinct and discrete states. Energy states may be called quantized if they do not transition from one to the other through intermediate forms. The cosmic pyramid is thus a metaphor for the build-up of the universe in a quantized manner, and shifts between time periods in the Mayan calendar amount to quantum jumps. From this perspective, synchronicities, such as independent ideas or sudden insights, are reflections of quantum jumps in the Tree of Life.

THE ACCELERATION OF TIME

The step-by-step activation of higher Underworlds with more rapid shifts between Days and Nights, and a twenty times higher rhythm with every new Underworld (see table 2.1), explains a phenomenon that has intrigued scientists and philosophers for several decades. This is the increased rate of novelty in the cosmic evolution scheme as we come closer to the present time. *New phenomena, in other words, seem to come into existence at an increasing rate as we approach the end of the Mayan calendar.*

Such a speed-up of the process of evolution is also evident in the mind-related evolution of Underworlds 5–8 (figure 2.5), which for most

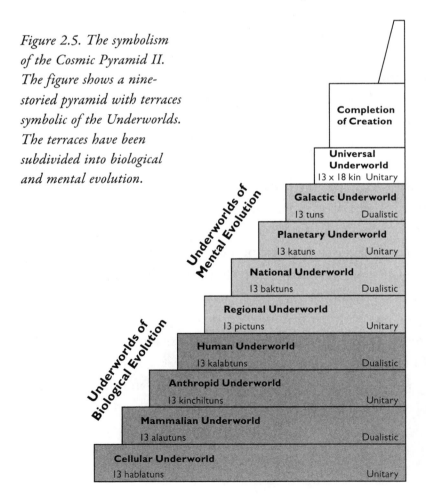

Figure 2.5. The symbolism of the Cosmic Pyramid II. The figure shows a nine-storied pyramid with terraces symbolic of the Underworlds. The terraces have been subdivided into biological and mental evolution.

Underworlds of Mental Evolution

Underworlds of Biological Evolution

Completion of Creation	
Universal Underworld 13 x 18 kin	Unitary
Galactic Underworld 13 tuns	Dualistic
Planetary Underworld 13 katuns	Unitary
National Underworld 13 baktuns	Dualistic
Regional Underworld 13 pictuns	Unitary
Human Underworld 13 kalabtuns	Dualistic
Anthropid Underworld 13 kinchiltuns	Unitary
Mammalian Underworld 13 alautuns	Dualistic
Cellular Underworld 13 hablatuns	Unitary

**Increase in the frequency of shifts between DAYS
and NIGHTS with every higher Underworld**

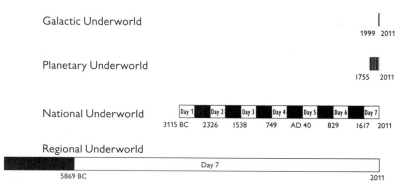

Figure 2.6. The acceleration of time. Each Underworld develops according to a rhythm of its own, defined by the duration of its Heavens. Since these durations become twenty times shorter every time a new higher Underworld is activated, Kairos shifts happen more frequently, more novelty is generated, and time is experienced as going faster.

people may provide the most illustrative example of the hastening of change. We may use the development of clothing fashions as an example of this acceleration of novelty. Among the Ice Age people in the Regional Underworld (figure 2.6), which was activated about one hundred thousand years ago, we know from archaeology that change was very slow, and there is no reason to believe that they were aware of any evolution of their lifestyle. Because of the slow rate of change, you could wear a mammoth fur your whole life without this going out of fashion. In the National Underworld, on the other hand, which was activated about five thousand years ago in 3115 BCE, and had a twenty times higher rhythm than the Regional, even as far back as Medieval Europe clothing fashions would among the nobility clearly change at a rate of about once every decade. In the Planetary Underworld, which began in CE 1755 and is the one that most readers of this book were born into, there came a point when the fashion houses launched four collections a year and most everyone noticed several changes in fashion colors on a yearly basis. (It is also incidentally because of the speed-up of time that occurred at the beginning of this Underworld that people became aware of the phenomenon of evolution in the late eighteenth century.) In the past 250 years, which has seen the industrialism of the Planetary Underworld, people have had a desire for everything that is "brand-new," and the speed of most kinds of change was much higher than previously. After the currently dominating Galactic Underworld began in 1999, most everyone is aware of another speed-up of time, at least on an inner plane, and more events of change seem to be compressed into shorter time periods. As a result, in certain countries stress and burn-out became noticeable health problems right when this Underworld began in 1999. The acceleration of time has become part of everyday life, either through the common experiences of time "disappearing" and "speeding up" or in people "not having enough" time. Yet, as we have now seen, the current speed-up has a prehistory and builds on stepwise increases in the rhythm of life of lower Underworlds.

Several metaphors have been used to describe the speed-up of time that takes place throughout the course of cosmic evolution. One such is Carl Sagan's "Cosmic Clock," in which the entire lifetime of the universe since the big bang is compressed into one year.[22] If on this "clock" the big bang took place on New Year's Day, then the Earth would have been formed in August, the first multicellular organisms would have emerged in November, and the first higher mammals on December 29. *Homo erectus* would have appeared at 10:48 p.m. on December 31, and the voyage of Columbus would have taken place one second before midnight on the last day of the year. Sagan's clock is very illustrative, but it does not explain why the speed of evolution accelerated.

The increased rhythm of change in the different Underworlds of the Mayan calendar may in fact be the first explanation for this cosmic clock. The Mayan calendar is unique among the calendars of the world to express a hierarchical structure of evolutionary progressions that develop at increasing speeds, and from this we may understand why time seems to be speeding up. It explains why there is a reality to the acceleration of time both in the short term of human life from which the examples above were taken and, as we will see later, in the long-term process of biological evolution. Sagan's Cosmic Clock, we may note, seems to indicate that evolution is driven by something other than random mutations that led its process to accelerate. As the cosmos, and we humans, climb the nine-storied pyramid, the different Underworlds are sequentially activated, one by one. Every time this happens processes developing with a twenty times higher rhythm are coming into play. It should be pointed out however that the Mayan stone pyramids were not actually built to scale with the rhythms of the different Underworlds. (As we can see from figure 2.6, this would not be technically possible.) The acceleration process is in fact much more dramatic than the architecture of these would indicate.

In reality, of course, it is not time in the sense of quantitative mechanical time that is speeding up (the speed of the revolution of the Earth around the Sun is not increasing). It is instead the rhythms of the evolutionary processes emanating from the Tree of Life that are proceeding at increasing rates as new Underworlds are sequentially activated. This incidentally highlights the difference between the two aspects of time. While Kairos is speeding up, Chronos, mechanical time, is unchanged. We may then wonder if the cosmos is approaching a change of a more fundamental nature. This is something we may ponder, as all of the Underworlds are relatively soon, at the same time, October 28, 2011, coming to completion.[23]

A PRELIMINARY TEST OF
THE TREE OF LIFE HYPOTHESIS

In chapter 1 I asserted that it is a testable hypothesis whether the recently discovered "Axis of Evil" in reality is the Tree of Life, since this would mean that the evolutionary pulses emanating from this axis would conform to the evolutionary scheme provided by the Mayan calendar. It may now, after we have acquainted ourselves with the Mayan calendar system, be appropriate to make a preliminary test of the hypothesis that the by modern science inappropriately labeled "Axis of Evil" in fact is the Tree of Life. If all of the Underworlds define evolutionary processes that can be described by the metaphor of going "from seed to mature fruit," we may start by studying what those seeds and fruits look like in the different Underworlds (table 2.5). By studying the starting points of each Underworld (table 2.2), it is clear that the planting of each seed means that a new relationship between life and its environment is created. In the Cellular Underworld of thirteen hablatuns beginning 16.4 billion years ago, this "seed" was then the organizational event of the big bang, where the first matter in the universe was formed. As the seventh Day of this Underworld began, a new and very significant organization of life in biological evolution emerged as the "fruit," and this was the higher, or eukaryotic, cells that emerged 1.2 billion years ago.[24]

Such eukaryotic cells are called "higher" because of their vastly increased complexity, and increased size, compared to lower, prokaryotic cells, such as bacteria. Their emergence was a necessary step for the continued evolution, since only eukaryotic cells (for the most part) can evolve into multicellular organisms. The first Underworld is called Cellular because at the beginning of its seventh Day, the higher cells emerged (as the metaphorical fruit of its process). Higher cells also, unlike their predecessors, often display behavior indicative of a primordial form of consciousness. *Paramecium,* for instance, the simple slipper animalcules, is a eukaryotic cell that may react to stimuli and move in the direction of food or light.

The emergence of these higher cells at the end of the first Underworld provided the foundation for the second Underworld, called the Mammalian. This second Underworld began 820 million years ago, which is also a reasonable estimate of the time when the first cluster-like multicellular organisms may have emerged, and these embarked a long process of increasing complexity. In this Underworld new classes of multicellular organisms developed with an alautun rhythm, twenty times higher than previously, and as it reached its seventh Day, 63 mil-

TABLE 2.5.

PRELIMINARY TEST OF THE TREE OF LIFE HYPOTHESIS

Underworld	Mayan Calendar (Beginning of 1st Day)	Modern Estimate	Mayan Calendar (Beginning of 7th Day)	Modern Estimate
Human	2.05 Myr	2.0 (humans)[a]	0.16	0.16 (*Homo sapiens*)[b]
Anthropoid	41 Myr	40 (monkeys)[c]	3.2	3.2 (*Australopithecus africanus*)[d]
Mammalian	820 Myr	800–900 (animals)[e]	63.1	65 (placental mammals)[f]
Cellular	16.4 Gyr	13.7 (big bang)[g]	1.26	1.2 (higher cells)[h]

[a]*Encyclopedia Britannica.*

[b]T. D. White, B. Asfaw, D. DeGusta et al., "Pleistocene *Homo sapiens* from Middle Awash, Ethiopia," *Nature* 423 (2003): 742–747, doi:10.1038/nature01669.

[c]Frank E. Poirier, *Understanding Human Evolution* (Englewood Cliffs, NJ: Prentice Hall, 1990), 103.

[d]Ibid.

[e]Philip Whitfield, *From So Simple a Beginning: The Book of Evolution* (New York: MacMillan, 1993) and Donald R. Prothero, *Evolution: What the Fossils Say and Why It Matters* (New York: Columbia University Press, 2007), 165. Whitfield gives 850 MYA and Prothero gives 800–900 MYA.

[f]Matthew Hedman, *The Age of Everything: How Science Explores the Past* (Chicago: University of Chicago Press, 2007), 125.

[g]M. Tegmark et al., "Cosmological Parameters from SDSS and WMAP," *Physical Review D* 69 (2004): 103501.

[h]Lynn Margulis and Karlene Schwartz, *Five Kingdoms,* 3rd ed. (New York: W. H. Freeman and Co., 1998).

lion years ago, its "fruit" came in the form of the emergence of higher mammals.[25] Mammals are clearly more intelligent than the reptiles and amphibians that preceded them and are also the only class of animals that displays behavior that is playful and, as pet-owners know, creative. This attests to the increased autonomy of the higher mammals in relation to their environment, which is a significant aspect of the evolution of species. Hence, as the seventh Day of the Mammalian Underworld dawned, a new plateau was reached, upon which monkeys and humans evolved.

From the time when the third Underworld began, 41 million years ago, fossils from the first monkeys have been found.[26] While humans often tend to look upon monkeys and apes as just another group of animals, these really relate to the environment in a way that is intermediate between other mammals and human beings. Among mammals, the use of tools is something that is typical of, but not completely exclusive to, monkeys and apes, and the ability to use them clearly reflects the new, more interactive relationship between organisms and their environment that developed in the Anthropoid Underworld. The first monkeys developed according to the 3.2 million-year-long kinchiltun rhythm of this Underworld to *Australopithecus africanus* at the beginning of its seventh Day. The latter were very advanced primates that distinguished themselves by their fully erect posture and bipedality, as evidenced by the famous 3.6 million-year-old footprints in Laetoli.[27] They represented a significant step toward the human being in that their arms were free for using tools.

On the basis of this the first members of the genus *Homo* appeared, approximately as the fourth Underworld—the Human Underworld—began 2.1 million years ago.[28] The first of these belonged to a species that goes by the name of *Homo habilis,* whose fossils were discovered in Central Africa, which is believed to be the ancestral home of all human beings. This species was the first to be included in the *Homo* category because there is anthropological evidence that they not only used, but, unlike the anthropoids, also made tools. Moreover, *Homo habilis* had a brain volume of 600 cm³, which was very significant in relation to its body size. The "seed" provided by this species is then developed by this Underworld through a sequence of pulses according to the kalabtun rhythm of 160,000 years, until its seventh Day began. There is today a consensus that *Homo sapiens,* which has a brain volume of 1,400 cm³, first appeared on the planet around 160,000 years ago,[29] which coincides with the beginning of the seventh Day of this Underworld.

Homo sapiens, in fact, not only completes the evolution of the fourth Underworld, but also completes the process of biological evolution as such, since the higher Underworlds (numbers 5–8) continue with the evolution of the mind of this particular species called *Homo sapiens* that we all belong to. Quite in contrast to what has been claimed by adherents of the Darwinist theory, the human being is at the top of the scheme of biological evolution. The matching of these four evolutionary processes from "seed" to "fruit" to the current estimates for the times of emergence of the respective species is, as we can see, very good, and if we restrict ourselves to looking at the "fruits," it is absolutely perfect. This means that the data are in very good agreement with the hypothesis that the Central Axis that emerged in the big bang is the Tree of Life, since the continued emergence of seeds and fruits of different species conforms very well with what would be expected from the evolutionary processes generated by the Tree of Life according to the Mayan calendar. I feel it is fair to say that this preliminary test of the Tree of Life hypothesis is a success.

We have seen that the modern dates are consistent with the view that each new Underworld in the Mayan calendar gives rise to an increased autonomy of the biological organisms versus their environment. This, in turn, supports the idea that an increased self-consciousness and an increasingly higher level of intelligence is developed by each Underworld and that these features have their origin in the Tree of Life. What we have found here is not only that evolution takes place according to a time plan, but also that this is expressed in the self-consciousness of biological organisms. Moreover, as higher Underworlds with increased rhythms are activated, an acceleration in intelligence takes place quite consistent with the metaphor of Sagan's Cosmic Clock.

The discovery of this time plan for evolution directly clashes with Darwinism, the theory that for essentially 150 years has been held by academics as the cornerstone of biological science (and often presented as a proven fact). In the Darwinist paradigm, biological evolution is not expected to follow a time plan. It is instead believed to be caused by random gene mutations that by definition—because of their randomness—occur at unpredictable points in time. In the theory presented here, the human being is instead the end result of four apparently directed, evolutionary processes, each generating an increasingly higher intelligence at an accelerated rhythm. Before we make a more detailed test of the new theory of biological evolution, we will turn our attention Darwinism in order to understand why it is in conflict with what has been presented here.

3

Basic Questions Regarding Biological Evolution

THE HISTORICAL BACKGROUND TO THE IDEA OF EVOLUTION

Theories about our origins are at the heart of what it means to be a human being, and whether we are aware of this or not, it is likely that the ideas we have about biological evolution directly determine how we perceive our purpose in life. If humans build their lives on their view of their origins, there is little wonder that the nature of this has generated heated debates. Our origins are crucial for our understanding of who we are and what we are here to do. As is natural for such a significant question, it has a long history. For most of the five thousand years of history, humans perceived themselves as having been created by "the gods" or God. Moreover, until the mid-eighteenth century, they were more inclined to say that life on Earth had devolved from an earlier Golden Age, or the Garden of Eden, than to consider the possibility that it had evolved to its current state from something more primitive. In many creation stories there is a point in the past where either the human beings distance themselves from the gods or the gods distance themselves from the human beings, which is then referred to as a Fall. Following this, people perceived the world as being essentially as it had been at the time of the Fall. Around the time of the scientific revolution in Europe in the early seventeenth century, most thinkers and the public alike abided by the view of the Bible and estimates of the age of the world that were derived from this. Famous among these age estimates is that made in CE 1654 by the Archbishop of Ardagh, who said the world had been created on October 23, 4004 BCE, an estimate that still today is essentially shared by so-called Young Earth Creationists.

In the mid-eighteenth century a change took place in the human mind (especially in the West) that strongly favored a rationalist approach to the world. This manifested in the philosophical and scientific movement of the Enlightenment. The first atheists emerged in the philosophical circles of France, and an anticlerical movement swept the world that among other things led to the separation of church and state in some countries. An aspect of this change in mentality was that philosophers and scientists came to embrace the idea of evolution rather than devolution. Initially, the notion of evolution was more of an abstract philosophical concept, such as Hegel's thesis-antithesis-synthesis evolutionary model or Schelling's view of nature as a series of stages of evolutionary processes by which spirit struggles toward consciousness of itself.

The short estimates of the age of the Earth based on the predominant reading of the Bible came to be increasingly recognized as inconsistent with the existing observations. In 1755 the German philosopher Immanuel Kant in *Universal Natural History and Theory of the Heavens* proposed that the world was million of years old. He also suggested that the solar system had evolved from a cloud of gas and that on a larger scale the Milky Way had the same origin. This immensely influential philosopher/scientist vastly expanded the time scale of existence and of course put in question what at the time seemed like a literal reading of the biblical account of our origin as a species. Kant also expanded the spatial framework of our existence well beyond the solar system and proposed the existence of nebulas (galaxies) as well as the possibility of life on other planets. An interesting example of how the idea of evolution was beginning to be entertained at the mid-eighteenth century is the shift in thinking of Carolus Linnaeus. Linnaeus had laid the foundation for systematic biology by introducing a classification system for the different species in his *Systema Naturae* (1735), in which he saw species as immutable creations of God. Yet in 1766 he changed his view and came to believe that a species could indeed also be changed into another.

In 1785 the Scotsman James Hutton published his main work on the formation of rock on sea floors by sedimentation, a process that would require immense periods of time. Regarding some of the species he found in such rocks, he concluded: "We find in natural history monuments which prove that those animals had long existed; and we thus procure a measure for the computation of a period of time extremely remote, though far from being precisely ascertained."[1] Following this, more and more evidence became apparent in the early nineteenth century, especially in the form of fossils, to indicate that the geological eras of our Earth spanned time periods in the range of Kant's estimate. From the study of fossils it then also became increasingly evident that most animals and plants that had lived on Earth no longer existed and that most existing organisms had appeared relatively recently. These facts required an explanation, and Jean-Baptiste Lamarck was the first notable scientist to propose a theory of biological evolution to account for these fossils. Like most of his contemporaries, including Charles Darwin[2] somewhat later, Lamarck believed that acquired properties could be inherited and so this was the basis of his theory of evolution. (An idea that has recently seen a revival in a branch of science called epigenetics.[3]) Lamarck also recognized what he saw as an inherent trend in evolution from the simple to the more complex, and in this his theory was different from what later would be proposed by Darwin, whose name is the one that more than any other has come to be associated with the concept of biological evolution. Since, however, there are many possible theories of biological evolution—Lamarck's theory was one and the Tree of Life theory presented in this book another—I feel it is important to call on Darwin's particular brand, Darwinism. At least to me, the idea of biological evolution cannot be scientifically disputed, but what is controversial in Darwin's interpretation of evolution is that it is said to proceed through accidents and without any direction. To him we can trace the philosophy of randomness that underlies current academic biology, and many other branches of science as well. Yet in our preliminary test of the Tree of Life hypothesis we found a few things that are inconsistent with this philosophy, namely that (1) evolutionary jumps do not occur at random points in time, (2) evolution has a direction toward higher intelligence and self-awareness, and (3) evolution is accelerating. Thus there are some very fundamental reasons to question Darwinism as an explanation for biological evolution.

After a long incubation period, Charles Darwin in 1859 published his ideas about biological evolution in *On the Origin of Species*. The book immediately became immensely influential and was reprinted seven times before its author passed away in 1882. It later spawned several related theories in other branches of science, perhaps especially in the humanities, and has had a profound effect on the emergence of political ideologies and how we see our lives in general. (In general parlance Darwin's name has become synonymous with the raw struggle for survival.) In the (next to) last words of his book, Darwin wrote:

> These laws taken in the largest sense, being Growth with Reproduction, Inheritance which is almost implied by reproduction; Variability from the indirect and direct action of the external conditions of life, and from use and disuse; a Ratio of Increase so high as to lead to a Struggle for Life, and as a consequence to Natural Selection, entailing Divergence of Character and the Extinction of less-improved forms. Thus, from the war of nature, from famine and death, the most exalted object which we are capable of conceiving, namely, the production of the higher animals, directly follows.[4]

This sums up his theory and is hardly any more uplifting for those hoping that there is a purpose to life than Steven Weinberg's words.

It should be acknowledged that even if Darwin was not the first researcher to propose the idea of biological evolution, he was nonetheless by far the most important scientist to have demonstrated that an evolution of the species indeed had occurred. He made this idea accepted both among the public and the scientific community, and it was a major contribution to our knowledge. To say this does not mean however that I agree with the explanation he presented. Unlike some of his contemporary biologists, Darwin completely rejected all forms of Platonism, according to which evolution would be guided by an "idea." Richard Owen, for instance, saw the body plans of the major phyla and other recurring structures in biological organisms as being developed much like a "crystallization" of cells along immutable immaterial ideas that he called "pre-determined or primal patterns." As I will attempt to demonstrate, this notion comes much closer to the true mechanism of evolution than Darwin's natural selection based on functionality.

Because of the absence of any purpose or divine guidance in Darwin's theory, the theory was naturally opposed, especially by Christians, and concerning the question of our origins a verbal war between science and religion began 150 years ago. While the scientific community increasingly sided with Darwinism, the opposing religious camp, at least in the West, mostly stood by the interpretation that the seven Days of creation in Genesis were Chronos days and that the Earth thus was no older than ten thousand years, a line of thinking that today would go by the name of Young Earth Creationism. On this basis a pseudoscience was developed that maintained that the dinosaurs had been contemporary with the humans within this ten-thousand-year time span. This line of thinking was based on an ambition to prove that the Bible is an infallible source of knowledge rather than on an unbiased search for the truth and it thus became an easy target for Darwinists in the scientific community. Since anyone respecting the ideal of rational empiricism in science will shun such disregard for facts, creationism got a bad name. Especially in the debates that followed, both sides tended to look upon evolution and creation as mutually exclusive. The conflict has, if anything, intensified in recent times after attacks by leading scientists, such as Oxford biologist Richard Dawkins[5] and members of the U.S. National Academy of Sciences,[6,7] on the idea that the universe has a creator. At the same time, highlighting the distance between the professional scientific community and the public, a current poll shows that as much as 48 percent of Americans are essentially Young Earth Creationists and only 13 percent believe in the Darwinist view that God has no part in evolution.[8]

As many as 30 percent of Americans however believe that there has been an evolution, which has been guided by God, but to date no serious scientific theory has supported this widespread intuition. The present book is meant to fill this gap by providing a solid scientific theory based on empirical evidence that the universe is dominated by a creative evolutionary process. This is not primarily motivated by a desire to broker peace, but, as we shall see, to tell the truth about the matter. In this new theory both sides in the protracted evolution/creation conflict will incidentally get to be right. The Darwinists will get to be right in the sense that there undeniably has been an evolution of biological species over billions of years leading up to the present time, but the biblical creationists will also get to be right in that evolution indeed follows a purposeful and nonrandom rhythm of seven Days and six Nights as described in Genesis. In this sense the new theory is inclusive and is part of a recontextualization of all of science in accordance with the commonsense anthropic principle.

DARWIN'S THEORY OF BIOLOGICAL EVOLUTION

As mentioned earlier, the formulation of Darwinism was partly prompted by the discovery of fossils that demonstrated that the organisms inhabiting this planet had undergone very dramatic changes over geological time spans. How could these changes be explained and, most important, what forces had been driving them? The answers that Darwin came up with were based on Malthus' theory about population growth and the observations of biological variation that he had made on his journey with the HMS *Beagle,* which led him to propose his particular model of evolution, which dispensed with a creator, a purpose, and a direction.

Charles Darwin believed that all organisms descended from common ancestors and that a variation in heritable traits, and a struggle for life among an offspring too great in number for all to survive, provided the basis for natural selection. This supposedly accounted for the survival of those species of animals and plants that were the most fit. In its original form his theory was really a combination of three basic ideas: (1) all species descend from a common ancestor, (2) random changes create variation leading to new species, and (3) the fittest genetic variants survive. Darwin believed that the individuals with the most favorable traits for the reproduction and survival of a species would outcompete less viable variants and, partly through geographic isolation, lead to a "descent with modification." The variation from which the viable species were selected was presumed to have emerged from *small, random* changes. He asserted that biological evolution took place through the effects of *slow, gradual* changes dispersed over time, without any major qualitative jumps, much like the breeding of domesticated plants and animals, which had been practiced for millennia. What determined the survival value of a trait was in his view how much it helped a species adapt to its environment. Hence, in Darwin's model, there was no role for a directed or purposeful evolution. Rather, alterations in the environment, including the influence of other species present, would favor the survival of the genetic variants best adapted to those circumstances. This is called natural selection.

In Darwin's lifetime the molecular background to genes was unknown (the term *genetics* was coined in 1905), and contrary to most accounts, he shared Lamarck's view that acquired properties could be inherited by the offspring, as we may see in the previous quote,

"Variability . . . from use and disuse." It was instead one of his followers, August Weissman, who changed his original theory and introduced the dogma that acquired properties could not be inherited,* which became the basis for the modern form of Darwinism. Unbeknownst to Darwin and thinkers at the major centers of learning, Gregor Mendel, a monk at the monastery of Brno in Moravia, had around 1860 performed experiments in his cloister garden that demonstrated that certain traits, such as the colors of flowers or the shapes of peas, were inherited independently of one another. This pointed to the existence of some kind of "packages" of inheritance that later came to be called genes. As the work of Mendel after his death was rediscovered around 1900, he became hailed as the father of genetics, and as Weissman's idea then seemed to be consistent with these new findings, genetics came to be embraced as the basis of Darwinism by the scientific community of the twentieth century. Later it was also discovered that genes were located on pairs of chromosomes in the cell nucleus.

Mendel's demonstration of independently inherited genes thus seemed to increase the plausibility of Darwin's theory, since it provided a mechanism that allowed for inherited biological traits to remain in the offspring. When James Watson and Francis Crick elucidated the structure of the DNA molecule (which was already known to hold genetic information) in 1953,⁹ a solid material basis for the existence of genes, and for Darwin's theory, seemed to have been discovered. The DNA, holding information about how the proteins are to be synthesized, had the structure of a long spiral chain (the double helix) located in the chromosomes and was made up of component molecules called nucleotides, which are symbolized by the letters A, G, C, and T. The translation of these nucleotide letters to amino acids in proteins became known as the genetic code, and the sum of all the genes the genome.

---

*Weismann advocated the germ plasm theory, which meant that (in a multicellular organism) inheritance takes place only through the germ cells: egg cells and sperm cells. This idea would seem to rule out the inheritance of acquired characteristics.

The proteins, it may be added, are large molecules that may be described as the workhorses of cellular metabolism and serve among many things to direct and facilitate all the chemical reactions in the cells. They are made up of twenty different amino acids and form large folded chains with specific three-dimensional structures that define their tasks.

These new findings meant the beginning of a vast revolution in biochemistry and molecular biology and, and in an amazing body of work in the 1950s the mechanisms of synthesis, replication, and transcription of the DNA molecule were elucidated and understood in a way that seemed fully consistent with Mendel's gene packages. It was found that messenger RNA was transcribed from the DNA and sent to the ribosomes, where the twenty amino acids by means of the different types of transfer RNA were brought together to build protein chains in a process like an assembly line (figure 3.1). Based on these findings the central dogma of biochemistry was coined: In the cell information flows from DNA via RNA to proteins.

It was also found that some agents, such as radiation and certain chemicals, were able to bring random DNA mutations about, but mutations could also occur spontaneously (without an external causative factor) with a certain frequency. These induced and spontaneous mutations in the DNA seemed to provide for exactly the kind of random genetic variation that Darwin had postulated, since according to the central dogma these mutations ultimately would be expressed in altered proteins that would result in visible (so-called phenotypic) changes in the traits of the organisms. It seemed as if this central dogma could provide a background for how the random changes in the traits that Darwin had envisioned could be produced through chemical alterations of the DNA sequence, by so-called mutations.

Even if it were not possible to know if mutation indeed was the mechanism that evolution had actually followed, at least two real-life cases seemed to support the idea of the survival of the fittest: (1) DNA mutations in bacteria and parasites would sometimes favor those genetic variants that were resistant to antibiotics, and (2) a mutation in

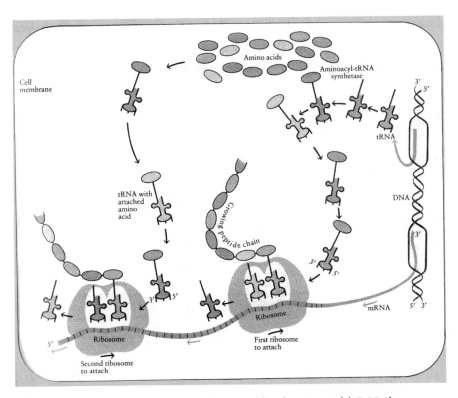

Figure 3.1. DNA and the central dogma of biochemistry. (1) DNA's structure is a double helix of paired strands. (2) Based on information in the DNA, the proteins, the actual agents or workhorses of cellular life, are synthesized. According to the central dogma of biochemistry, information flows from DNA through RNA to instruct the cell which proteins are to be synthesized. Coding sequences of DNA transcribe their information to messenger RNA, which, in a process similar to an assembly line, combines amino acids carried by transfer RNA to form long protein chains. With permission from Helena Curtis and Sue Barnes, Biology, *5th ed. (New York: Worth, 1989).*

human hemoglobin-generating red blood sickle cells provided protection against malaria and hence the sickle-cell gene was favored in areas where malaria was highly prevalent. Yet another example was the black peppered moth, which had outcompeted the previously dominating white variety as industrial England became sooted, apparently because its color provided camouflage from discovery by predators. A few such examples have been found indicative of natural selection, where under certain circumstances individuals with specific DNA mutations survived because they were more fit compared to their ancestral forms. Yet in retrospect I think it is fair to say that the examples are extremely few in number and very marginal for a theory that often is presented to the public as proven to be true and that has been held to be so certain that it has been allowed to change the thinking of our civilization. Scientists are far from lazy, and if natural selection indeed was the predominant mechanism behind biological evolution hundreds of examples would by now be known.

Nonetheless, these examples were sufficient for the scientific community to accept the theory that came to be called Neo-Darwinism, after it had been integrated with genetics and the central dogma of biochemistry. Moreover, in support of Darwin's notion of a fundamental unity of all life, common descent, the genetic code furnished by the DNA was found in all forms of cellular life studied and was verified by protein sequencing, showing that all forms of life on the planet very likely are related. Thus it did not appear as if the different species had been created by God "after their kind" as stated in Genesis. The victory of Neo-Darwinism seemed total, at least within the scientific community, and in universities and medical colleges all over the world, Neo-Darwinism came to be presented as a proven fact.

RANDOM MUTATIONS ARE RARELY BENIGN

Despite this apparent consensus within the academic community in support of Neo-Darwinism, many people outside this community have remained unconvinced, and to most it has probably seemed counterintuitive that human existence would merely be an unplanned whim of nature. The most questionable aspect of the Neo-Darwinist model of biological evolution is its apparent lack of direction, since the driving force behind it is random mutations. In practice such mutations have been demonstrated to bring about only very small changes within an already existing species. Random mutations have thus never been shown to initiate the evolutionary leaps necessary to bring about a new species. I myself have worked twenty years at some of the leading scientific laboratories in the world, whose focus has been to develop methods to identify the substances in the human environment that cause random mutations (so-called mutagens). The reason for this effort has been that such substances are not only harmful to the offspring, but also tend to cause cancer and several other unwanted effects on the ecosystem and ourselves. Thus agencies such as WHO/IARC, USEPA, and others see it as their responsibility to regulate and minimize public exposure to substances that cause random mutations. From such a perspective it then seems quite counterintuitive to suggest, as is a critical part of the Neo-Darwinist theory, that these very substances would be the driving factors behind evolutionary change in a positive sense.

It seems worthwhile to point out that among the few examples that have been found where mutations have increased the survival value of an organism (and, as mentioned above, this does happen), these mutations nonetheless do not drive evolution forward. They merely benefit survival. For instance, in the absence of the external threat of malaria, sickle-cell anemia is clearly a step backward that decreases the health of the affected individuals. While it has been shown that in certain instances single mutations, such as those giving rise to sickle-cell anemia, may benefit the survival of a species, they fall very much short of producing anything like a new healthy and more evolved species. The emergence of new species in fact usually requires large-scale *synchronized* transformations of several distinct physiological mechanisms, metabolic pathways, and cell components, and such synchronized changes have not been shown to be inducible by random mutations.

It is actually not very surprising that it has not been possible to create new species by random mutations, since even one single functional and benign protein would be extremely unlikely to emerge by mere chance. Consider a protein of normal size that is constituted by 100 amino acids. Since there are 20 different amino acids that could all be in a given position, the probability that a protein with an exact given amino acid sequence should emerge randomly is $(1/20)^{100}$—or about 1 chance in 10^{130}—an exceedingly small figure. Because many amino acids are in positions where they have no effect on the function of a protein, the probability of a such emerging by chance is somewhat higher, 1 in 10^{65}, where the latter however still is an exceedingly large number that is equal to the total number of atoms in our galaxy.[10] This is the probability of *one* new functional protein emerging through random processes. In the cell, several different proteins however often exert their functions as part of a group in which each one of them is critically needed, and the probability of all of these proteins emerging by accident is truly negligible. We run into absurdly low probabilities—similar to when we discussed the basic constants of nature. The likelihood for the structure of a single protein, not to mention a whole organism such as the human with 20,488 genes,[11] to emerge by accident is simply implausibly low, roughly 1 in $(10^{65})^{20,488}$. The famous British astronomer Fred Hoyle said this better with a metaphor: The probability of life emerging by accident is less than that a tornado sweeping through a junkyard might assemble a Boeing 747.

BIOLOGICAL CHANGE IS RARELY GRADUAL

Another significant objection that has been raised many times in the past against Darwin's theory is that the fossil record does not for the most part substantiate his prediction that evolutionary change proceeds through many small steps, mutation by mutation, and is therefore slow and gradual. The many examples of sudden emergences in the fossil record were evident already in Darwin's own lifetime, including to some extent to himself, but no tenable explanation has been produced for these sudden jumps leading to new species. While some organisms remain essentially unchanged over billions of years, and some undergo slow gradual change,[12] more often than not evolutionary changes happen through leaps in which a new species suddenly appears in the fossil record with many novelties and with few or no ancestral forms. To illustrate the standard model of evolution, biologists use phylogenetic trees, where new classes of organisms branch off from their original lineages (figure 3.2).

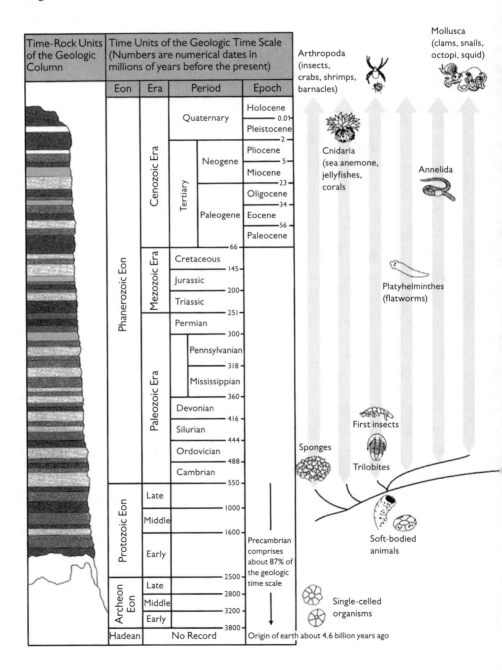

To exemplify such qualitative leaps in biological evolution, consider the transition from water to land by certain animal species, such as from fishes to amphibians. This requires not only the transformation of gills into lungs and fins into legs, but concurrent changes in the skin and the reproductive system and so on. Or take, for instance, the transformation of a land-based mammal into a whale, which is generally believed to have taken place. That transition would require not

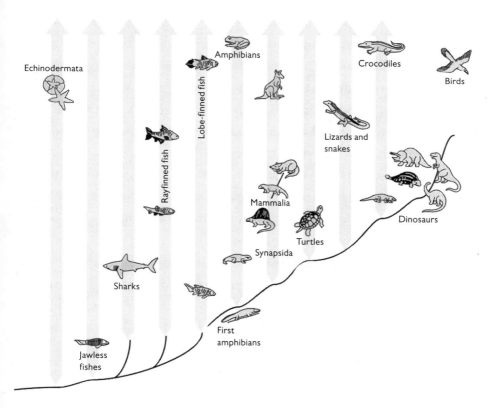

Figure 3.2. Phylogenetic tree of the evolution of different classes of organisms. This phylogenetic tree shows the current view of biological evolution on Earth, where different species have branched off during the past 4.5 billion years. Used with permission from Donald R. Prothero, Evolution: What the Fossils Say and Why It Matters *(New York: Columbia University Press, 2007).*

only that legs be turned into fins, but also that a respiratory system be fully developed, allowing the cetacean to be under water for long periods of time with a temperature-regulating system to sustain cold water. Without trying to detail the necessary transformations needed in this or the many other sudden leaps in the emergences of new species, the point to realize is that all of the different physiological changes would have needed to be *synchronized* in order for the new species to be viable. None of these changes in isolation would have any survival value unless they happened in synchrony with the others. Because of the need for synchronized changes in many different organ systems, these could not possibly have been a result of slow, gradual change according to the Darwinist model. Even if Darwinists sometimes have very lively imaginations and produce ideas that allow reality to fit their theory, I think it is fair to say that Darwinism lacks any explanation as to how synchronized changes take place in several different organ systems. Yet such synchronized changes seem absolutely necessary for several evolutionary transitions that indeed have taken place. This points to a need for a synchronizing factor to explain evolutionary leaps, and any hint of such a factor is lacking in Darwinist theory.

I would like to suggest that the Tree of Life brings synchronization about in biological evolution. If, however, the Tree of Life were to serve as a synchronizing factor in evolutionary transformations, we would, because of its cosmic nature, expect these to take place simultaneously in many different species. We would find examples of so-called convergent evolution, where animals from different lineages simultaneously develop similar traits. This is indeed exactly what we find, since in the emergence of classes of different organisms there seem to be specific points in time when a high number of species of the most diverse kinds suddenly leap into existence with some new biological trait, often preceded by few, if any, preliminary forms. An example of such a point in time is the so-called Cambrian explosion, which took place 570 to 540 million years ago and suddenly produced a vast array of species of diverse kinds with shells. Recent studies have shown that such broad emergences of many different new species appear with a clear periodicity[13] in the fossil record, which is something we will come back to as we present a more detailed test of the Tree of Life hypothesis in the next chapter.

Such cyclical bursts of new species clearly do not fit with the Darwinist theory and its basic idea of slow, gradual change. Evolution instead seems to occur through periodic "quantum jumps"—sudden dramatic transformations of organisms in accordance with a specific rhythm. It is also noteworthy that it has not been possible to match this periodicity with any changes in the Earthly environment, and thus the major transformations in evolution do not seem to occur by natural selection of successful adaptations to environmental changes or catastrophes. Regardless, there is no reason to believe that an environmental change or catastrophe of any kind would be able to synchronistically bring about the many parallel changes in an organism that the transition to land would require. This suggests that the cyclicity of the appearance of new species is a rhythm inherent in the process of evolution itself.

QUANTUM JUMPS IN BIOLOGICAL EVOLUTION AS EVIDENCE OF INTELLIGENT DESIGN

If it could be demonstrated that any complex organ existed which could not possibly have been formed by numerous, successive, slight modifications, my theory would absolutely break down.

CHARLES DARWIN

The inability of Darwinism to explain sudden transformations in the fossil record makes it into a theory that does not really explain the observed data in a way that you would normally require of a scientific theory, and some facts especially speak against it: (1) The extreme improbability that new functional proteins or organelles (cellular organs) would emerge by random processes; (2) the necessity for parallel processes to be synchronized for the transition to land, to sea; or to air; and (3) the lack of survival value for an organism to have a nonfunctional organ, such as half a wing. All of the mentioned aspects illustrate the failure of Darwinism to explain the rapid transformation of biological species at several points in the fossil record.

Among the features mentioned above, we still need to touch upon item 3, which is an argument that was first raised against Darwinism in the nineteenth century, but has not yet been countered. The argument is as follows: Birds, for instance, could not have emerged from a Darwinist process of selection of the fittest from reptiles, simply because there is no survival advantage for a reptile to have only a small nonfunctional wing that does not allow it to fly. On the contrary, a reptile with a nonfunctional half wing would be less fit for survival, since it would have to expend energy on making the half wing and drag this around even though it had no use. If this is true, a wing could never evolve by natural selection, since any species with a quarter or a half wing would be at a disadvantage and selected against. In several species similar situations would occur for many different organs that, as mentioned above, would only have selective value together with synchronized changes in other organs. A retina, for example, would not be meaningful if there is no brain to process visual information or a lens to focus it.

Such arguments apply also to cell structures and biochemical processes, and in the early 1990s a group of scientists came forth as proponents of the idea that biological organisms must have emerged by intelligent design. Michael Behe, who is one of the leading scientists advocating this idea, has expanded the same basic argument raised above regarding the reptilian wing to apply also to the realm of cell biology, pointing out that if in Darwin's day the complexity of biochemical mechanisms had been known, his theory would probably not have been accepted or even proposed. An interesting example of such complexity raised by Behe is that of flagella in bacteria that cannot possibly have resulted from thousands of random mutations in the several different proteins that they are made from. This is again because it is only the presence of a complete and functional organelle made from a range of different proteins that would favor a selective value for the bacteria. There would thus be no survival value for a cell to produce one of these particular proteins if it was not part of a functional whole. For a cell to produce a nonfunctional flagellum would simply mean that it would unnecessarily expend energy and disturb its own metabolism. This organelle, and the way it is built by the cell, is in fact exceedingly complex; it could only have emerged through the coordination of several synchronized parallel processes and not possibly through a linear sequence of random nondirected mutations. Behe sees the flagellum as an example of "irreducible complexity" and supports this with strong arguments based on the complex mechanism with which this organelle is formed in the cell.[14]

This may seem intellectually valid and innocuous enough, but the idea of intelligent design provoked a very strong reaction among the scientific establishment in the United States, and a large number of its professional organizations have denounced it. Very little scientific criticism was part of this denouncement, and the intransigent conflict is to a large extent one of ideologies. The debate between the two sides has also to a large extent been confused by the fact that both have tended to counterpose evolution and intelligent design as if they were mutually exclusive. In reality no serious scientist today can question the reality of evolution, and the real issue is whether the evolution we observe is created by random mutations such as in the Darwinist model or by cosmic design. While the fears of liberal-minded scientists that the ideas of religious special interest groups, sometimes embracing Young Earth Creationism, will gain traction to some extent are understandable, the irony is that the response of the professional scientific community to this initiative exhibits exactly the same kind of fundamentalism that it supposedly claims to want to counter.

The response of the established scientific community also illustrates the fear of the immense consequences it would have if it became more widely known that Darwinism is a false theory. Confidence in academic science would plummet with wide-ranging ideological, religious, and political ramifications. Many applied sciences such as medicine, pharmacology, genetic engineering, agriculture, and other related multibillion-dollar industries would have to rework their approaches against much resistance because of the enormous amounts of human energy that have been invested in the support of Darwinism. Rather than opening up this Pandora's box by admitting its inconsistencies, the U.S. National Academy of Sciences and other professional scientific organizations have elected to present a hard front against the idea of intelligent design and pretend that the Darwinism model is a proven fact that no professional biologist can question.

There is however one point in the criticism that has been raised against the idea of intelligent design that I find valid, that its adherents have not been able to present a coherent theory as an alternative to the Darwinist explanation of biological evolution. I think it needs to be recognized that it would not be easy for educators to show students that the Darwinist theory has been proven wrong without presenting an alternative to it. Personally, I also doubt that it is possible to single out specific phenomena, such as organelles or organs, as designed if in fact the whole universe is intelligently designed. Isolating specific phenomena as designed could give rise to the idea that there may be intervening acts of creation—something I dispute. Hence I will talk only about the universe as intelligently designed rather than focusing on special aspects of the biological organisms.

BIOLOGICAL EVOLUTION INCREASES THE COMPLEXITY OF ORGANISMS

The last shortcoming in Darwin's theory that will be brought up here is that it provides no explanation for the fact that overall, in the course of evolution, biological organisms have evolved toward higher complexity and become increasingly more conscious and intelligent. Thus the oldest organisms discovered on this planet are akin to blue-green algae, from which a few billion years later emerged higher eukaryotic cells, which in turn evolved not only into plants, but also through worms to shellfish to fishes to reptiles and to mammals, stepwise to human beings, which to the naked eye looks like an increase in complexity.

To Darwinists the absence of an explanation to this increase in complexity is not seen as a problem, since the existence of a direction of evolution is simply denied. The late Stephen Jay Gould sums up the Darwinist position well: "There is no progress in evolution. The fact of evolutionary change through time doesn't represent progress as we know it. Progress is not inevitable. Much of evolution is downwards in terms of morphological complexity, rather than upward. We're not marching toward some greater thing."[15] The same message, worded differently, is repeated by most biological textbooks, giving the impression that there must be some evidence to support this notion or that it is more than someone's personal philosophy of life. To most people a theory that does not explain the evolution toward more complex and advanced forms simply seems very counterintuitive. Since we actually observe an evolution toward higher complexity, it would seem like a failure of a scientific theory not to be able to explain this. In contrast, in our preliminary test of the Tree of Life hypothesis it became evident that this actually accounts for a biological evolution toward higher intelligence and complexity. In figures 4.4, 4.5, and 5.1 it will also become clear that each Underworld has a distinct role in increasing the complexity of life and that the complexity increase that each one of them brings about can be quantified. The Underworlds thus each drive biological evolution to increasingly higher levels of complexity, and as has been hinted, the end result of this directed evolution is *Homo sapiens,* who in turn became the starting point for a mental evolution leading to even higher degrees of complexity.

The Tree of Life hypothesis, however, is not the only one that acknowledges that biological evolution shows a trend toward increasing complexity. It has been proposed that the universe and its biological organisms have an inherent tendency to self-organize.[16] Philosophically, the concept of "self"-organization is quite questionable, and we may feel compelled to ask: If the big bang was self-organized, what is then meant by "self"? Also, even though we obviously observe a trend toward higher complexity throughout cosmic evolution, there seems to be no experimental evidence that this would be the result of any "self-organization." On the contrary, if we mix together the different components of a cell, they will fail to "self-organize" into a cell, and the proponents of this theory have not explained why this is. More profoundly, we may ask: If biological organisms emerge through an inherent tendency to "self-organize" in this universe, why then do the quantum jumps in evolution adhere to the Mayan calendar? It seems that phenomena that some ascribe to self-organization would better credited to Halo-organization, since this allows us to see them as part of a larger coherent context of biological evolution. Thus the complexity increase in biological evolution follows certain patterns and is not just random. It is only against the background of Mayan cosmology, and its charting of quantized Kairos time, that we are able to identify these patterns in the complexity increase. It is from this perspective that we will come to understand some important things about the direction and purpose of creation. In the next chapter we will embark on this exploration.

4

The Tree of Life Hypothesis

THE CELLULAR UNDERWORLD

The time has now come to more rigorously test the hypothesis that the Cosmic Tree of Life is the synchronizing factor behind biological evolution, which is what this chapter will be devoted to. This primarily entails testing how well significant emergences of new species in biological evolution match the shift points between Days and Nights in the four lowest of the Underworlds of the Mayan calendar. We will start with the first, and most basic, of these, the Cellular Underworld, which is the only one that spans the entire age of the universe. The name given to each Underworld indicates that we are looking upon them primarily from the perspective of the emergence of biological life (the ancient Mayan names are not known). The first Underworld has thus been named for cells, which it develops into their higher, eukaryotic forms. Table 4.1 describes the evolution of life in the Cellular Underworld, as much as is known to us at least. The seven Days and six Nights of this Underworld each lasts for a hablatun (1.28×10^9 tun).

TABLE 4.1. THE EVOLUTION OF LIFE IN
THE CELLULAR UNDERWORLD

Heaven	Beginning (Gyr)	Major Events
Day 1	16.4	Big bang (16.5 Gyr, WMAP flat universe[a])[b] Milky Way (14.5 +/- 2.8 Gyr)[c]
Day 2	13.9	Big bang (13.7 Gyr, WMAP favored estimate[a])[b]
Day 3	11.3	
Day 4	8.8	
Day 5	6.3	Formation of star systems with Earth-like planets (6.4 Gyr)[d]
Night 5	5.0	Earth (>4.6 Gyr),[e] meteor bombardment giving water to Earth (4.6–3.9 Gyr)[f]
Day 6	3.78	Prokaryotic cells (3.8 Gyr)[g]
Night 6	2.5	Oxygen (2.5 Gyr), extinction of anaerobes[h]
Day 7	1.26	Eukaryotic cells with cilia and flagella (1.2 Gyr)

[a]Note that the two dates for the big bang are alternative estimates of the age of the universe based on different assumptions.

[b]M. Tegmark et al., "Cosmological Parameters from SDSS and WMAP," *Physical Review D* 69 (2004): 103501.

[c]N. Dauphas, "The U/Th Production Ratio and the Age of the Milky Way from Meteorites and Galactic Halo Stars," *Nature* 435 (2005): 1203–05.

[d]Charles H. Lineweaver, "An Estimate of the Age Distribution of Terrestrial Planets in the Universe: Quantifying Metallicity as a Selection Effect," *Icarus* 151 (2001): 307–13, http://arxiv.org/pdf/astro-ph/0012399v2.

[e]Matthew Hedman, *The Age of Everything: How Science Explores the Past* (Chicago: University of Chicago Press, 2007), 160.

[f]Peter D. Ward and Donald Brownlee, *Rare Earth: Why Complex Life Is Uncommon in the Universe* (New York: Springer, 2003), 123.

[g]Ward and Brownlee, *Rare Earth*, 97.

[h]Ward and Brownlee, *Rare Earth*, 177.

Generally speaking, in any given Underworld, the dating of phenomena is more difficult and uncertain the further back in time we go. This is especially true for the age of the universe, which according to the Tree of Life hypothesis and the Mayan calendar would be expected to be 16.4 billion years (Gyr) old. While this is close to the commonly used rough estimate of 15 Gyr, there are today many scientists who favor a slightly younger universe of 13.7 billion years. Clearly, if uncontestable measurements indicated that the big bang took place only 5 billion years ago or, alternatively, as much as 25 billion years ago, it would invalidate the Mayan calendar estimate of the time for the emergence and activation of the Cosmic Tree of Life. Considering however that its deviation even from the low value of 13.7 Gyr is relatively small (20 percent) and that the measurements involved are very difficult and the results derived from them have fluctuated considerably over the years, I feel there is no reason to reject this slightly higher estimate of the age of the universe. The point to realize is that such age estimates of the universe directly depend on many assumptions about its fundamental nature that, especially considering the current shaky state of cosmology, may very possibly change. In a recent study based on WMAP measurements, the authors settled for an age of the universe of 13.7 Gyr,[1] but include a range of estimates up to 16.5 Gyr, based on different assumptions as to the nature of the universe. Moreover, a recent estimate of the age of the Milky Way, our local galaxy, published in *Nature,* was given as 14.5 ± 3.8 Gyr,[2] and this age obviously cannot be higher than that of the universe itself. Even if I do not think that it can be concluded that the Mayan calendar estimate of 16.4 billion years is in perfect agreement with modern estimates, I feel it is definitely in the range of them. I should add that I personally think that the Mayan estimate of the age of the universe is the correct one, and some reasons for this will be supplied as we go along, but I also feel that the reader has the right to know to what extent this deviates from those that are currently the most favored by cosmologists.

Although the Cellular Underworld is the longest of the nine Underworlds, there is for most of its duration no information available to us about the evolution of life corresponding with it. Most traces from the universe's early years seem to have disappeared completely, including members from the first generation of stars. For this reason, any discussion of the evolution of life in the first 10 Gyr after the big bang will be highly speculative. Yet although it is not provable or disprovable, I entertain the view that primitive forms of cellular life, and at the very least failed attempts at creating such, have existed around stars other than our Sun, since the very beginning of the Cellular Underworld, that is to say from the big bang.

Yet the only life we know about is that from within our own star system, which presumably was formed during Day 5 of the Cellular Underworld. Star systems are believed to form as extrastellar gases condense into increasingly larger masses, which form a rapidly rotating disc of matter. From the core of this a star is then lit up as a shining object, while solid matter around it accretes into planets and asteroids. One of the consequences of this model of star formation, which also been supported by photographs,[3] is that the formation of planetary systems from matter accreted around stars may be assumed to be very common. This model is believed to be valid for our own solar system, and it is thought that, because of the intense radiation of heat from the Sun as it lit up, most of the lighter elements necessary for cellular life evaporated from the inner planets more than 4.6 billion years ago. A currently relatively well-accepted theory maintains however that the water in the seas of the Earth, and much of its current organic material, came back to it through a bombardment by meteors and comets during its early existence, 4.6 to 3.9 billion years ago.[4] We may still see scars from this period of bombardment in the craters of neighboring celestial bodies, such as the Moon, Mercury, and Mars. This bombardment took place because early in the history of the solar system the orbits of many comets and planetoids were not stable. Gravitational effects from the Sun and the major planets then influenced these asteroids and brought them out of orbit, until they eventually collided with the major celestial bodies, planets, and their moons. Only after this initial period, when stray asteroids bombarded the planets, did the solar system gain its present relative stability in which meteor impacts are rare.

During this period of bombardment, the Earth suffered what we would probably liken to hellish conditions, and each impact raised the temperature to levels that prevented oceans from forming and the conditions suitable for life from stabilizing. Because of the high temperatures created by these impacts, the Earth may have been barren and inhospitable to life until about 3.9 Gyr ago.[5] Then, only 100 million years later, the first prokaryotic cells seem to have appeared on Earth—apparently as soon as the temperature was low enough to allow this. Prokaryotic cells are either bacteria or belong to a group called archeae, and the oldest of these that have been discovered may be related to the present cyanobacteria, which are also known as blue-green algae. When contemplating this scenario where life seems to have emerged as soon as the physicochemical conditions allowed for it, it is hard to escape the feeling that "life" was somehow already waiting to materialize on our planet in the form of these first simple prokaryotic cells

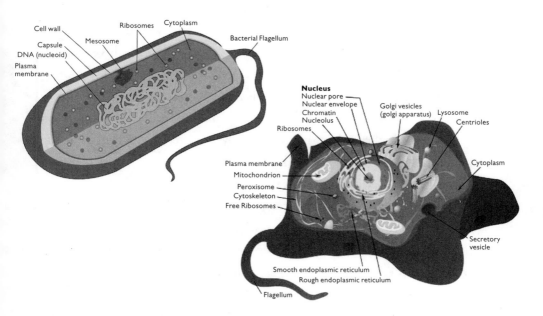

Figure 4.1. Prokaryotic and eukaryotic cells. Cells belong to two basic types, (a) prokaryotic and (b) eukaryotic. Bacteria and archeae are prokaryotic cells, which lack an inner differentiation into organelles (cellular organs). Eukaryotic cells are much larger and have several different organelles. All multicellular organisms, as well as the types of unicellular organisms that are called protists, are made up of eukaryotic cells.

(figure 4.1a). In fact it is believed to have done so exactly at the beginning of the sixth Day of the Cellular Underworld, 3.8 Gyr ago.[6]

Such a viewpoint, where life is "waiting to happen" at the "right time," certainly warrants a new definition of life, one that will be discussed in chapter 7, and it is probably already now evident that the approach we are taking here is quite different from the conventional. The entire Cellular Underworld is here believed to have propelled the emergence of life in a universal and galactic context in a process going back to the big bang. I do not, in other words, believe that the first cells suddenly emerged merely because of a series of random chemical reactions on Earth 3.8 Gyr ago. This does not seem reasonable, given what we know about the extremely low likelihood of even one functional protein in such a cell to emerge by accident, as well as the barren and inhospitable surface of Earth at the time. Instead, I propose that cellular Halos (or Spherical Space-Time Organizers) that had existed since the inception of the universe were only waiting to manifest life. From such a perspective the very rapid emergence of cells on Earth, immediately as the temperature became sufficiently low to allow for this, seems to be less of an enigma, since this may be seen to have been brought about by a step in the Cellular Underworld, namely the sixth Day.

The first higher eukaryotic cells finally appeared at the beginning of the seventh Day. Eukaryotic cells are different from prokaryotic in that they have a nucleus (*karyos* literally means "nucleus") and also a number of other organelles—cellular organs—like mitochondria, liposomes, Golgi apparatuses, chloroplasts, and lysosomes (figure 4.1b). Some of these organelles are known to have emerged by a process called endosymbiosis (*endo* means "inner," *sym* means "together," and *bios* means "life"), which means that microbes were engulfed by a larger host cell and then came to perform specialized functions in the latter. The first eukaryotic cells may have emerged some 1.5 billion years ago, or even earlier, but it was about 1.2 billion years ago that they underwent rapid diversification with sexual reproduction. This time frame also saw the development of cilia and flagella,[7] which are small feelers that help cells move and interact "consciously" with their surroundings. As we shall see, these organelles play a very significant role in the evolution of multicellular organisms, and without them an organism does not really qualify as conscious.

In terms of testing the Tree of Life hypothesis, we may then conclude that the times of emergence of the two major types of cells, prokaryotes and eukaryotes, fit perfectly well with the expectation of steps forward taking place at the very beginnings of Days 6 and 7. Even if there is some uncertainty regarding the dating of the big bang, the picture is consistent with the view that new forms of life are propelled into existence by a wave movement of pulses (Days) emanating from the Cosmic Tree of Life, which was activated at the birth of the universe. We can also see that the Nights, in contrast, meant the extinction of some forms of life. In Night 5 for instance, any possible early temporary cellular settlers on Earth must have gone extinct during the meteor bombardment. During Night 6 the many anaerobic life forms that are believed to have emerged during Day 6 met the same fate, since the oxygen atmosphere that emerged during this Night was toxic to them. Yet despite these destructive aspects, Nights 5 and 6 also directly paved the way for the new prokaryotic and eukaryotic life forms that appeared on Earth during Days 6 and 7, respectively, by providing water and oxygen to the Earth. As mentioned earlier, Nights often serve such preparatory roles. Thus in the wave movement of an Underworld—any Underworld—the Days represent Light and movement forward in the evolution of life, whereas Nights are "resting" periods of Darkness. Yet it is in those very resting periods that the coming steps forward are prepared. From such considerations emerges a rationally based view, where the evolution of life is a reflection of a wave movement going back to the birth of the universe.

THE WAVE MOVEMENT
OF MULTICELLULAR SPECIATION
The Mammalian Underworld

We will now step up to the second Underworld on the Cosmic Pyramid, the Mammalian Underworld, which, with a twenty times higher rhythm of evolution than the Cellular, develops the first multicellular organisms into the higher mammals. This Underworld, like the others, has been given its name from its end result. In this Underworld, evolution is propelled by thirteen alautun periods, each of which lasts 63.1 million years, meaning that its wave movement spans an era of 13 x 63.1 = 820 million years. Its evolutionary process builds on the fundament provided by the single eukaryotic cells that were created in the Cellular Underworld, and from this it generates a new level of life with a much higher complexity: multicellular life. Since there is considerably more information regarding biological evolution in this second Underworld, it is critical for actually testing the Tree of Life hypothesis.

Traditionally, paleontologists have studied the evolution of animals and plants by classifying the past several hundred million years into different geological eras. Such eras have been given names like Cambrian, Triassic, Devonian, and Mississippian based mainly on the types of fossils characteristic of them. This classification however has not always been made on clearly defined criteria, but rather on fairly subjective or geographically defined characteristics of the animal life found in different geological layers. The negative aspect of this is that it has created a certain bias and probably even blurred an understanding of the course of important events during biological evolution. It is natural that the use of such era names without a strictly defined systematic basis predisposes the student to look at the emergence of new species as rather chaotic. The approach here is instead based on eras defined by the different Days and Nights of the Mammalian Underworld. This may then also create a bias, but with the significant difference that this bias is clearly defined.

However, it is not just the somewhat arbitrary subdivision of geological eras that may have contributed to the impression that biological evolution is a chaotic and random process. It is also the particular traits and biological phenomena that have been chosen for study, those considered the most relevant in the study of evolution. Paleontologists naturally have an interest in bones and skeletons, since those are often all that can be found in fossil form, but if one is interested in the direction of evolution, the study of such may be misleading. Thus, as I hinted at in the preliminary test of the Tree of Life hypothesis in chapter 2, we will here study the evolution of multicellular organisms from the perspective of the increased brain power, and self-awareness, of different classes of biological organisms.

Multicellular organisms (metazoa) originate almost exclusively from eukaryotic cells, which as mentioned are cells made up of different compartments with separate functions, so-called organelles. A few multicellular organisms are formed by prokaryotes, but complex animals and plants are invariably based on eukaryotic cells. The background to the complexity of multicellular organisms is that the types and numbers of organelles in eukaryotes can be varied so that different cell types emerge. Certain cells, like muscle cells, for instance, have a high number of mitochondria that produce the energy for their activities. Red blood cells lack a nucleus and are filled with hemoglobin, whereas white blood cells specialize in producing immunoglobulins for the protection of the whole organism against foreign substances. Because of their different internal structure, the different cells can be classified into different types, and through the integration of these in specialized organs a high capacity for certain functions may emerge.

A key feature of many multicellular organisms, and certainly all animals with a higher form of intelligence, is their basically bilateral body plans. In fact, all higher animals, including all vertebrates, are essentially symmetrical, and in addition to a left and a right side, which essentially are mirror images of each other, they have a front-back and a dorsal-ventral polarity, giving them the appearance of having emerged from a three-dimensional coordinate system. Only lower animals with a very simple nervous system, such as mollusks, starfish, jellyfish, or coral, have body plans that are asymmetrical or radial (figure 4.2). The left-right symmetry in higher animals is very marked for the eyes and the central nervous system, and a polarized and symmetrical system seems in fact to be a prerequisite for higher "intelligence." (Organs that primarily have metabolic functions, such as the liver or the intestines, may however sometimes be asymmetrically placed.) It may also be noted that the entire universe, through its Central Axis, is symmetrical, and possibly polarized into sections of galaxies with different handedness, which are mirror images of each other. This suggests that maybe the body plans of bilateral multicellular animals, including that of human beings, is a reflection, or creation, "in the image" of the Cosmic Tree of Life, and that intelligence, including that of our ancestral species, is derived from resonance with this polarized universe.

It is interesting to note that within the basic symmetry of intelligent species, there is a certain degree of asymmetry in the mirror images. We know that even if the two brain halves of an animal are mirror images of one another, they are not perfectly symmetrical. In humans many studies have verified that the two brain halves have different mental functions, and the same is true for the biological functions of many other animals.[8] The left-right polarity of the brains of higher animals may be at the heart of the phenomenon of intelligence for the very reason that it reflects an overall polarization of the universe. Regardless, the Darwinist paradigm fails to explain why the body plans of higher animals are built around an invisible three-dimensional coordinate system, and why among them there is a total dominance of bilateral forms, something we will return to in the last chapter of this book.

The development of classes of animals in the direction of increasingly

Figure 4.2. Asymmetric, radial, and bilateral body plans. Multicellular organisms display a wide variety of body plans, which are most basically subdivided into asymmetrical, radial, and bilateral. All higher animals are bilateral, meaning they have a left and a right side that essentially are mirror images of each other. Courtesy of Bengt Sundin.

more developed bilateral central nervous systems is shown in table 4.2 as a function of the Days and Nights of the Mammalian Underworld. Before discussing the details of this table, a cautionary note about methodology is due. As I pointed out earlier, the dating of phenomena usually becomes more difficult the further back in time we go. This is true also for the Mammalian Underworld. Generally speaking, the dating of fossil materials is fraught with many difficulties, and while the dates may fall within certain ranges, they vary from scholar to scholar and fluctuate over time. Moreover, all paleontologists wonder how complete the fossil record actually is, and this is obviously something that is very difficult to know, since many species may not fossilize well or be preserved at all.

TABLE 4.2.

THE EMERGENCE OF MAJOR CLASSES OF BIOLOGICAL ORGANISMS IN THE MAMMALIAN UNDERWORLD

Heaven (Day or Night)	Beginning (Myr)	Classes of Organisms[a]
Day 1	820.3	First clusters of cells (850)
	757.2	
Day 2	694.1	Ediacaran Hills fauna (680)
	631.0	
Day 3	567.9	Cambrian explosion: trilobites, ammonites, mollusks (570)
	504.8	
Day 4	441.7	Fishes (440)
Night 4	378.6	Protogymnosperms (372)[b]
Day 5	315.5	Reptiles (300)
Night 5	252.4	Gymnosperms (275)[b]
Day 6	189.3	Mammals (190)
Night 6	126.2	Angiosperms (120)[b]
Day 7	63.1	Placental mammals (65)

[a]Datings within parentheses are from Philip Whitfield, *From So Simple a Beginning: The Book of Evolution* (New York: MacMillan, 1993) or from

[b]K. J. Niklas, B. H. Tiffney, and A. H. Knoll, "Patterns in Vascular Land Plant Diversification: An Analysis at the Species Level," in Valentine, James W. ed. *Phanerozoic Diversity Patterns: Profiles in Macroevolution*, 97–128 (Princeton, NJ: Princeton University Press, 1985).

The famous fish species *Latimeria* may be used to illustrate the inherent difficulty in charting fossils. This species of fish is believed to represent a crucial step between the fishes and the amphibians and reptiles, having four feet that later made the transition to land. *Latimeria* is found in the fossil record from about 400 to 70 million years ago,[9] but was for a long time believed to have gone extinct at the later time point. Thus when a fishing boat in the Indian Ocean in 1938 caught a living member of this species of fish, it produced somewhat of a sensation. Even if we may guess that it became less common 70 million years ago, the conclusion is nonetheless that it must have existed for these last 70 million years without leaving any trace in the fossil record. This shows the difficulties associated with proving any time-based theory about biological evolution. Thus it is natural that even different experts in paleoichthyology (the study of ancient fishes) may have varying opinions about whether the first fishes emerged 450, 440, or 430 million years ago. Sometimes expert datings may vary even more widely, especially when you go far back in time. Moreover, sometimes there is uncertainty because of the difficulty in defining exactly what is a fish.

Since I am not a trained paleontologist I have to rely on the work of others when it comes to datings. In order not to be criticized for choosing datings of fossils that perfectly fit my theory (in this vast literature it could easily be done) I have chosen to limit myself to merely one popularized (Darwinist) description of evolution, Philip Whitfield's *From So Simple a Beginning: The Book of Evolution* (see table 4.2). If the reader consults other sources, he or she will likely find that the datings are somewhat different, and the serious student is advised to do so in order to get a sense of the variation in the work of different scholars.

In table 4.2, when we look at the estimated times of emergence of significant classes of animals in the various Heavens, we can see that a class of species with a more developed bilateral central nervous system emerges very close to the beginning of every Day. The multicellular animals in Days 1 and 2 are faded and put within parentheses because the datings of the organisms quoted in Whitfield were not based on actual findings of multicellular organisms, but on commonly made speculations. The lack of preceding forms to the species generated by the Cambrian explosion was recognized as problematic even in Darwin's time. The absence of fossils from multicellular organisms early in the Mammalian Underworld is a problem for any theory about biological evolution, including the one that I am proposing here. The dates for the classes of organisms at the beginning of Days 1 and 2 should then merely be regarded as guesstimates, although it is generally agreed among paleontologists that such ancestral species must have existed and emerged around 800–900 million years ago.[10] My own hypothesis is that close to the beginning of Days 1 and 2, 820 and 694 million years ago, respectively, two significant steps in the evolution of multicellular organisms took place in the form of two pulses of primitive bilateral, but very thin, organisms that preceded those emerging in the Cambrian.

The Cambrian "explosion" of shelled animals has always been a problem for Darwinism, since this theory would assume that gradually formed predecessors to those that emerged at this time should have been found. In the new theory of biological evolution proposed here, it is instead expected that a marked change, or even an "explosion," indeed *should* happen every time a new Day begins, and the Cambrian explosion at the beginning of Day 3, from which time the oldest fossilized multicellular organisms are found, is perfectly consistent with this. Today many scholars would classify the species from the beginning of the third Day as belonging to the Ediacaran fauna and only recognize those from around 540 million years ago as part of the Cambrian. It should be pointed out that the term *explosion,* which has been used since the nineteenth century, is somewhat misleading. From the perspective of the Mayan calendar, a Day, such as the third Day, is more to be regarded as a *wave of novelty,* with the first new species with distinctly new characteristics turning up at the beginning of the Day. As the Day progresses, more and more novelty emerges until the wave crests at the midpoint of the Day. Then, as the following Night approaches, novelty continues to be created, but at an increasingly slower rate.

During the *third* Day of the Mammalian Underworld, essentially coinciding with the Cambrian era, a wide diversity of species emerged that produced a high number of fossils in the form of trilobites, ammonites, mollusks, and jellyfish, as well as species from all the phyla that currently exist. While it is likely that the introduction of shells in this era is the reason that these fossils have been preserved, we should notice that these shells also offered a necessary protection for more advanced nervous systems. Eyes are another novelty that came with the third Day, and they provided added sensory input and required increased brain power for processing. The fourth Day subsequently gave rise to the first fishes, marine organisms with a cranium and vertebrae, which both represented new steps for the protection of a central nervous system, providing the groundwork for further evolution. The fifth Day gave rise to the complete transition to land by the reptiles, while the sixth and seventh Days produced the lower and higher mammals, respectively, organisms that have brains that are markedly lateralized (whose halves have separate functions). The brains of the higher mammals of the seventh Day also display a much-expanded folding of the cortex.

This progression shows that if we focus on the development of a nervous system in increasingly bilateral organisms, then the evolution of species no longer appears as random. Instead, an increasingly more advanced nervous system is developed Day by Day, from seed to mature fruit. By focusing on this factor, what otherwise could present an immensely complex picture with the most diverse kinds of species is simplified and found to be consistent with the pulses forward during the Days of this particular Underworld. I may need to add however that I am not asserting that the beginning of Days were the only points in time when new species emerged. Even though the beginning of Days meant significant novelties, individual species and even classes of animals could emerge at many other time points in the progression as well. Yet when it comes to steps forward in the evolution of a bilateral central nervous system, it seems that those were concentrated at the beginning of Days, and this pertains especially to the emergence of the head as a distinct body part.

The scheme shown in table 4.2 was published in 2001 in my book *The Mayan Calendar*,[11] although I did not in that book include the dates for the emergence of the major types of plants. The hypothesis was presented then that the rhythm of evolution of multicellular species is dominated by an alautun rhythm of 63.1 million years. What adds drama to the assertion made then is that later, in 2005, Rohde and Muller published a much-noted article in *Nature*[12] that discussed the largest paleontological collection (the Sepkopski compendium)[13] of marine species that has ever been analyzed. The study included about 18,000 genera, and the purpose of the study was to identify possible cyclic periodicities in the emergence of new species. What the authors then found was a statistically highly significant periodicity of 62 ± 3 million years for the emergence of new species (see figure 4.3), which was later confirmed by an independent statistical analysis.[14] The authors of the *Nature* article discussed various ideas as to what might have generated this periodicity, such as different kinds of catastrophes or environmental changes, but found none that seemed to match the clear 62-million-year rhythm. So far the only explanation that matches the data seems to be the one I published several years ago based on the Mayan calendar. It should be recognized that my explanation was not constructed "after the fact," as so often happens, but was an a priori hypothesis that is there for anyone willing to examine it. It is generally agreed that in science, the verification of such predictions carry a considerably higher weight than explanations made afterward. In this case, however, when the prediction threatens the very foundations of Darwinism and by implication modern science, the existence of such a prediction is more likely to be ignored to the extent possible by established science. Regardless, the publication of the results of this large-scale study in *Nature* and the fact that those results had been predicted from the Tree of Life hypothesis mean that the hypothesis cannot be pronounced unsound unless the large-scale study for some reason proves to lack validity, which there is currently no reason to suspect.

This creates a rather curious situation. Because of their worldviews,

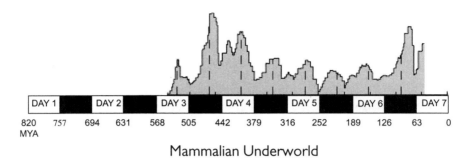

Mammalian Underworld

Figure 4.3. The Mayan calendar and the rhythm of diversification of marine species. The emergence of new marine species is highly correlated with the alautun rhythm of 63 million years, and the numbers of them show increases as new Days and Nights of the Mammalian Underworld begin. Data adapted from Robert A. Rohde and Richard A. Muller, "Cycles in Fossil Diversity," Nature 434 (2005): 208–10.

and a strong pressure to conform, many professional scientists summarily reject creationism in all of its forms as a kind of superstition that can never be verified by observations or experiments. In this case, however, based on the Tree of Life hypothesis, an excellent quantitative prediction was made, whereas Darwinists have made no predictions at all. And why would they? After all, the existence of such a cyclical diversification of species is not consistent with the Darwinist notion that evolution is generated by slow, gradual, random changes. Thus the verification by Rohde and Muller, using the largest paleontological collection ever studied, of my a priori hypothesis that biological evolution follows the rhythm of the Mayan calendar, indeed provides very strong support for the Tree of Life hypothesis and, in equal amount, disqualifies Darwinism.

It should be noted here that the emergence of new classes of animals seems to be induced by the Days, whereas the emergence of major groups of plants seems to occur close to the beginning of Nights (table 4.2). It is the combination of the two that creates the alautun rhythm. That new classes of plants essentially seem to emerge in the Nights means that they prepare the way for the classes of animals that follow in the coming Days. A typical example is the transition to land of the reptiles in Day 5 that had been prepared for by the protogymnosperm plants appearing on land in Night 4. Another example is that the emergence of the higher mammals in Day 7 was prepared for by the angiosperms during Night 6, which provided the necessary grass fields that have fed the browsing herds of mammalian herbivores ever since. Still today the angiosperms dominate the flora of Earth and make up all the flowers, fruits, vegetables, and grains, which are the feedstock of the mammals, including ourselves. Among the many necessary conditions for the emergence of intelligent life on the surface of the Earth, one is obviously that there be plants at the bottom of the food chain for animals to eat. Without plants, the transition to land by animals would not have been possible.

Hence, both Days and Nights generate new species, but of a somewhat different nature. While the Days generate increasingly intelligent animals, the Nights create plant species that may feed them. I would like to reiterate that in general the Days create more of a movement forward, whereas the Nights are periods of rest. This is consistent with the comparison that we may make between animals and plants, since plants that emerge during Nights are stationary, consistent with the resting character of those eras, whereas animals that emerge during the Days are mobile, consistent with the general movement forward during those eras.

In the Mammalian Underworld, the number of different cell types making up a species is an interesting indicator of the increasing complexity of life. Figure 4.4 shows an estimate of the increase in the number of cell types throughout the Mammalian Underworld. While the beginning of the curve is very uncertain, what is interesting about this curve is that it appears to be plateauing as it approaches the present time. This gives the impression that it is the purpose of the Mammalian Underworld to develop its particular form of complexity to a certain level, which then serves as a platform for the next Underworld of evolution. It should be

**Complexity increase in the Mammalian Underworld:
Number of cell types in different Heavens**

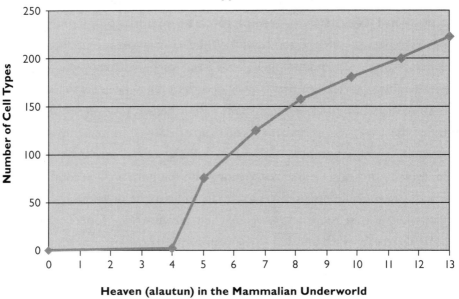

Figure 4.4. The increase in the number of different cell types in the Mammalian Underworld. The increased complexity of multicellular organisms during the course of the Mammalian Underworld is demonstrated in the increase of the number of cell types. Adapted from J. W. Valentine et al., "Morphological Complexity Increase in Metazoans," Paleobiology *20 (1994): 131–42.*

pointed out that the curve for the number of cell types does not include different types of nerve cells, since the subclassification of those is somewhat ambiguous.[15] The number of cell types in humans is however commonly estimated to be around 260.*

In terms of testing the Tree of Life hypothesis, we have now found evidence that classes of species with increasingly more advanced central nervous systems tend to appear at the beginning of Days of the

---

*In humans, 260 is a common estimate of the number of cell types (see for instance Ken Howard, "Interview with Stuart Kauffman," www.yuksel.org/e/guest/kauffman.htm). Yet several classifications are disputable and many would more cautiously say that it is between 200 and 300.

Mammalian Underworld. The complexity increase resulting from the number of cell types also seems to be a function of the progression of this Underworld. We have found evidence that the emergence of plants is a function of the beginning of Nights, and that Night phases prepare for the next evolutionary step among the animals. Furthermore, the work of Rohde and Muller has verified the a priori hypothesis that evolution in the Mammalian Underworld is a function of the alautun rhythm.

THE ANTHROPOID AND HUMAN UNDERWORLDS

The Mammalian Underworld did not result directly in the emergence of human beings with their highly developed ability to manipulate their environment. Instead, the higher, placental mammals, which emerged at the beginning of Day 7 of the Mammalian Underworld, were simple insectivores presumably akin to present-day shrews or moles. Human beings came about only through the influence of two additional Underworlds, each increasing the vertical component and size of the brains, and hence also adding a new level of intelligence to mammals, until *Homo sapiens* is produced at the seventh Day of the Human Underworld. I have called these two Underworlds the Anthropoid (previously called Familial, table 4.3) and the Human (previously called Tribal, table 4.4), again based on their end results. They develop according to the kinchiltun (3.2 million years) and kalabtun (160,000 years) rhythms, respectively, and span 41 million and 2.05 million years. The first of these, the Anthropoid Underworld, shifted the nervous system from essentially horizontal to vertical and served to develop monkeys into fully erect bipedal species that were able to use tools. The nonhuman primates and monkeys, which evolved in this Underworld, most certainly displayed a behavior with more control of their surroundings than any previous class of animal.

TABLE 4.3.

THE EVOLUTION OF IMPORTANT PRIMATE SPECIES
IN THE ANTHROPOID UNDERWORLD

Heaven (Day or Night)	Beginning (Myr)	Anthropoid Species[a]
Day 1	41	First monkeys (40)
Day 2	35	*Aegyptopithecus* (35)
Day 3	29	
Day 4	22	*Dryopithecus* (*Proconsul*, 20)
Night 4		
Day 5	16	*Kenyapithecus wickeri* (14)[b]
Night 5		
Day 6	9.6	*Australopithecus afarensis* (3.6)[c]
Night 6		
Day 7	3.2	*Australopithecus africanus* (3.0)

[a]Datings from Frank E. Poirier, *Understanding Human Evolution* (Englewood Cliffs, NJ: Prentice Hall, 1990).

[b]M. L. McCrossin et al., "Fossil Evidence for the Origins of Terrestriality among Old World Higher Primates," in *Primate Locomotion: Recent Advances*, ed. Elizabeth Strasser, John Fleagle, Alfred L. Rosenberger, and Henry McHenry, 353–96 (New York: Plenum Press, 1998).

[c]Note that this date does not really fit in Day 6 and that *Australopithecus afarensis* is here only assumed to be older than the fossils that have been found from it.

The exact reconstruction of the evolution of anthropoids, summarized in table 4.3, is fairly difficult, since fossils are lacking. What is known is that the first monkeys emerged at the beginning of this Underworld, shortly before South America and Africa drifted apart. In Burma, fossils from monkeys have been discovered that are about 40 million years old,[16] following which the New World and the Old World monkeys came to develop along separate lines. *Proconsul*, which is known from about 18 million years ago, seems to have shared many characteristics with the modern chimpanzee and is believed to belong to the same lineage. The scarcity of fossils from anthropoids would make it pointless to try to argue that the emergence of their different types appear close to the beginning of the Days. *Proconsul,* or *Dryopithecus* as its Latin name reads, may however have been typical of the anthropoids during the midphase of this Underworld, approximately 27 to 13 million years ago. Following this, there is a big gap in time between *Dryopithecus* and *Australopithecus afarensis* in which the fossil record is incomplete, although some chimpanzee-like ancestors of ours are assumed to have emerged then.

Fossils from a more evolved genus, the australopithecines, with clearly developed bipedality, have been found from the last 4 million years of the Anthropoid Underworld. The first fossil from this genus is the famous Lucy, a specimen of *Australopithecus afarensis* dated to about 3.18 million years ago that was discovered by Donald Johanson. This relatively intact skeleton (about 40 percent preserved) finally settled the issues of what came first, the erect posture or the increase in brain volume, in favor of the former. Translated to the terminology we are using here, this means that the Anthropoid Underworld developed the erect posture upon which the Human Underworld increased the brain volume. The first footprints from an erect *Australopithecus* in the volcanic ash of Laetoli originate from about the same time.[17] The French paleoanthropologist Yves Coppens[18] has labeled this *Australopithecus afarensis,* pre-*Australopithecus,* since it does not appear to have had the same stable erect posture as its follower, *Australopithecus africanus.* Consequently, it has been placed as a Day 6 species in the progression from seed to mature fruit. Yet no fossils have actually been found of this from such an early time.

Even though it is impossible to judge whether the fossil data are consistent or inconsistent with the Days and the Nights of the Anthropoid Underworld, the "seed" and the "fruit" in an evolution that decisively shifted the central nervous system from horizontal to vertical have nonetheless been found. The Anthropoid Underworld thus brings about a directed evolution toward bipedality and an erect posture. This transformation, incidentally, could hardly have taken place according to a Darwinian process of natural selection of the fittest, since two-legged animals move much more slowly than four-legged and any intermediate form would have had to have been even slower. Yet an erect posture is an important evolutionary step toward human beings, which means an improved overview of the environment and liberated hands that can be used for manipulating tools. From the perspective of the Tree of Life hypothesis, with its emphasis on development in three dimensions, an erect species seems more clearly to have been created in the image of an axis at the center of the universe.

There is not, however, any direct evidence that australopithecines used tools, and we will just have to assume that they began to use tools as their hands became free to do so. Their brain-to-body weight ratios were higher than in present-day chimpanzees, who sometimes use tools. An ability to use tools would indicate that the anthropoids were more autonomous in relation to their environment and hence had a greater potential for intelligence, compared to other placental mammals. Concurrent with these changes there seems to have been a development of the forebrains of these animals. Unfortunately, the fossil record for the anthropoids is too scant to allow us to diagram the transition to either an erect posture or increased intelligence that was brought about by this Underworld. Yet the Anthropoid Underworld clearly represents a distinct level of evolution between that of the mammals and that of the human beings.

When we begin to study the evolution of members of the genus *Homo* (table 4.4), we are on much firmer ground, as more fossils have been found from the Human Underworld. The definition of human beings generally accepted in anthropology is that we are organisms that not only use tools, but are also able to *make* tools. The first organisms that were able to do this were members of the species *Homo habilis,* which appeared slightly more than 2 million years ago. This species was discovered by the Leakeys and coworkers.[19] These differed from the australopithecines not only because of their larger brain and brain-to-body ratio, but also because their skulls show evidence of Broca's area, which is known to play a certain role in the use of speech.

TABLE 4.4. EVOLUTION OF THE GENUS *HOMO* IN THE HUMAN UNDERWORLD

Heaven (Day or Night)	Beginning (Myr)	Hominid Species[a]
Day 1	2.05	*Homo habilis* (2 MYA)
Day 2	1.76	Early *Homo erectus* (1.8)
Day 3	1.44	
Day 4	1.12	Late *Homo erectus*
Day 5	0.80	*Homo antecessor* (0.8),[b] use of fire (0.8)
Day 6	0.48	Archaic *Homo sapiens* (0.4)
Day 7	0.16	*Homo sapiens* (0.15)

[a]Datings from Steve Parker, *The Dawn of Man,* consulting ed. Michael Day (New York: Crescent Books, 1992).

[b]J. L. Arsuaga, I. Martínez, C. Lorenzo et al., "The Human Cranial Remains from Gran Dolina Lower Pleistocene Site (Sierra de Atapuerca, Spain)," *Journal of Human Evolution* 37 (1999): 431–57.

Shortly afterward, about 1.6 million years ago, *Homo erectus* emerged with a noticeably larger brain volume. Over the next million years, *Homo erectus* is believed to have spread from Central Africa to large parts of the Old World, and more recent examples of *Homo erectus* have been found in Java and China that have considerably larger brain volumes than the earlier African variants. Thus, *Homo erectus* was not an identical subspecies during the whole time of its existence. Instead, this species seems to have evolved through several stages—presumably corresponding to the different

Days of this Underworld. From about 300,000 to 400,000 years ago we find the first representatives of what has been called the "archaic" *Homo sapiens,* and finally about 160,000 years ago there appeared representatives of *Homo sapiens* proper.[20] We find a near-perfect concordance of the emergence of this highly significant organism, *Homo sapiens,* and the beginning of the seventh Day of the Human Underworld, 158,000 years ago.

While the modern human, biologically speaking, was almost fully developed 160,000 years ago, for all we know it was only during the Regional Underworld, starting some 100,000 years ago, that a diversified production of complex tools, and presumably a diversified language, began. It was in the Regional Underworld that humans started to represent their environment and themselves artistically, thus attesting to an emerging awareness of themselves and their mental abilities. Ultimately, the first four Underworlds served to develop organisms that biologically were prepared for the emergence of the human mind and spirit in the fifth Underworld.

The discussion of the Anthropoid and Human Underworlds has here been rather limited, and the few points raised are relevant to the proposal of a new theory of biological evolution. The evolution of the central nervous system and intelligence in the two Underworlds is an important aspect of this discussion. The Anthropoid Underworld made the central nervous system vertical and led to the use of tools, and the Human Underworld generated a tremendously rapid increase in brain size and ushered in tool production. For the Human Underworld, we can construct a diagram of increased brain size, since the number of fossils and their quality is much higher than from the Anthropoid. Figure 4.5 shows the brain volume of various species classified as *Homo,* as a function of the Human Underworld.[21] Again, as in figure 4.4, we have an indicator of how a form of biological complexity is developed by a certain Underworld, which seems to plateau toward the end. The bump shortly before the present time actually comes from the emergence of the Neanderthals, who are believed to have had a brain volume somewhat larger than ours. In the Human Underworld the increase in complexity is thus manifested in the form of an increase in brain volume.

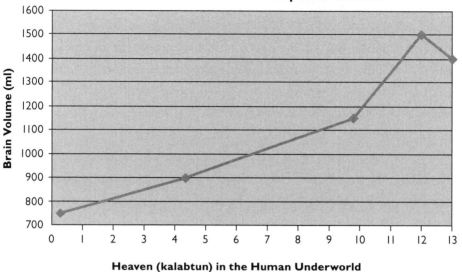

Figure 4.5. Increase in brain volume in different species of the genus Homo *in the Human Underworld. The increase in the brain volume is proportional to the number of neurons in the brain, which in turn gives rise to a thousandfold increase in the number of synapses, resulting in a vast increase in brain complexity. Adapted from data at www.wsu.edu.*

THE PERIODIC SYSTEM OF BIOLOGICAL EVOLUTION

Having now described the four lowest Underworlds on the Cosmic Pyramid, it is time to summarize our findings and look at the results in a unified context. I have called this context the periodic system of biological evolution, which is literally what its name says, a *periodic* system of biological evolution. This is a part of the entire periodic system of evolution, which summarizes evolution in all of the nine Underworlds, and which we saw another part of in table 2.4. Table 4.5 is limited to the four Underworlds that bring about biological evolution. As we have now climbed to the fourth level of the Cosmic Pyramid, the rhythm of evolution has increased tremendously. We have gone from the hablatun rhythm of the Cellular Underworld to the kalabtun rhythm of the Human Underworld, which means that the rate of biological evolution has increased some 160,000 times. It should now be clear that it is not correct to look upon biological evolution as a single process leading along a straight line from blue-green algae to *Homo sapiens*. Instead, each Underworld is generated by a wave movement that has a unique rhythm and unique characteristics, as well as a mechanism of creation and a specific purpose that distinguishes it from the others.

Before we study more of the differences between the various Underworlds in chapters 6 and 7 we will look for commonalities in the patterns of their unfolding. The periodic system of biological evolution (table 4.5) is set up so that relevant organisms emerging in the particular Days and Nights of the different Underworlds are placed in the same columns. This facilitates comparison of the progressions in the various Underworlds even if their rhythms are widely different. As pointed out earlier each Underworld is a creation of its own, although the Underworlds build upon one another and are also part of a total creation scheme. In this scheme, the organisms become conscious and responsive to changes in the environment in the Cellular Underworld; they start to take initiative and interact consciously with the environment in the Mammalian; they use tools in the Anthropoid; and they make tools to manipulate their environment in the Human. *Each Underworld thus carries an increased level of self-awareness of the biological organisms,* and from this what we call intelligence emerges. We may think of each of these Underworlds as having a special purpose in the overall creation scheme and that each Underworld generates a specific level of intelligence in the biological organisms. Ultimately, then, the various characteristics of the biological organisms generated by the different Underworlds, such as the ability to manipulate the environment, process information, and communicate, require some level of self-consciousness, a critical phenomenon that is increased as the Cosmic Pyramid is climbed. As part of this climb we can also see that biological evolution in its totality has a direction, as the different species in the

TABLE 4.5. THE PERIODIC SYSTEM OF EVOLUTION FOR THE FOUR LOWEST UNDERWORLDS

Ruling Quality	Cellular Underworld[a] of 13 hablatuns	Mammalian Underworld[b] of 13 alautuns	Anthropoid Underworld[b] of 13 kinchiltuns	Human Underworld[b] of 13 kalabtuns
1st Heaven is Day 1 **Sowing**	16.4–15.1	820–757 First Multicellulars?	41–38 First monkeys	2.05–1.90 *Homo habilis*
2nd Heaven is Night 1	15.1–13.9	757–694	38–35	1.90–1.74
3rd Heaven is Day 2 **Germination**	13.9–12.6	694–631 Early Multicellulars?	35–32 *Aegyptopithecus*	1.74–1.58 Early *Homo erectus*
4th Heaven is Night 2 **Reaction**	12.6–11.4	631–568	32–28	1.58–1.42
5th Heaven is Day 3 **Sprouting**	11.4–10.1	568–505 Ediacaran, Trilobites	28–25	1.42–1.26
6th Heaven is Night 3	10.1–8.8	505–442	25–22	1.26–1.11
7th Heaven is Day 4 **Proliferation**	8.8–7.6	442–379 Fishes	22–19	1.11–0.95 Late *Homo erectus*
8th Heaven is Night 4	7.6–6.4	379–316	19–16	0.95–0.79
9th Heaven is Day 5 **Budding**	6.4–5.1	316–252 Reptiles	16–13 *Kenyapithecus wickeri*	0.79–0.63 *Homo antecessor*
10th Heaven is Night 5 **Destruction**	5.1–3.9 Meteor bombardment, water	252–189 Perm-Tri Extinction Gymnosperms	13–9.6	0.63–0.47
11th Heaven is Day 6 **Flowering**	3.9–2.6 Prokaryotic cells	189–126 Early mammals	10–6.4 *Australopithecus afarensis?*	0.47–0.32 Archaic *Homo sapiens*
12th Heaven is Night 6	2.6–1.3 End of anaerobes, oxygen	126–63 Angiosperms	6.4–3.2	0.32–0.16
13th Heaven is Day 7 **Fruition**	1.3–0.0 Higher eukaryotic cells	63–0.0 Higher placental mammals	3.2–0.0 *Australopithecus africanus*	0.16–0.0 *Homo sapiens*

various Underworlds are brought to an increased mastery of their environments as they increasingly reflect the Tree of Life.

As was initially true also for Mendeleyev's periodic system of the elements, my periodic system has gaps, with periods we cannot fill in with certainty. Moreover, the very existence of the periodic system of the elements, we must note, prompted the search for the missing elements that would fill its gaps, since it helped chemists anticipate their properties and to know where to look for them. This was further facilitated in the twentieth century when quantum theory was developed to explain the periodic system of the elements. Similarly, we may also gain important information about what may be in the gaps by comparing the species in the different columns and rows of the periodic system of biological evolution. Such a lack of fossils especially pertains to the first several Days and Nights of the Cellular Underworld, which I suspect will never be filled by any real organisms, since the fossils would be some 10 billion years old—simple protocells at some Population II star system. The early Mammalian Underworld also contains gaps, but these may quite possibly come to be filled. The same may be true for the gaps in the Anthropoid Underworld. Generally, the empirical verifications are less certain in the early parts of all Underworlds, but overall despite its gaps the periodic system of biological evolution presented here seems to be quite consistent and contain enough detail for us to be able to base some conclusions on it.

By studying commonalities between specific Days and Nights, we may gain insights regarding the pattern of transformation common to all the Underworlds in their processes from seed to fruit. We will start our study of the system by comparing the manifestations during Days 6 and 7, which are the final phases of the Underworlds. In table 4.5, the manifestations emerging sharply at the beginning of Day 7 invariably are those that it is natural to refer to as the highest manifestations of significant evolutionary trends. (This is true whether or not you use the Mayan calendar as a reference.) At the beginning of Day 7 of the four lowest Underworlds, we have the higher cells, the higher mammals, *Australopithecus africanus,* and *Homo sapiens,* respectively, all of which represent clear completions of the different processes. Similarly, in Day 6 we see something that we may refer to as a protoform of these highest expressions; prokaryotes, lower mammals, *Australopithecus afarensis*, and archaic *Homo sapiens*, respectively, which in all cases are quite developed organisms, yet they seem to be one step away from the final expression of that particular evolutionary trend. In the Mayan calendar, Days 6 and 7 symbolically manifest the flower and the fruit, respectively, of the particular evolutionary processes, and in the columns of Days 6 and 7 we see the concrete manifestations of these symbols.

There are also start-up phases for the Underworlds, which take place during Days 1 and 2, when species with fundamentally new relationships to the environment emerge. (For these simple beginnings less information is available.) Following this is a series of three Days—Days 3, 4, and 5—that may be looked upon as a midphase of the process from seed to mature fruit. Each of these midphase Days has characteristics of its own in the overall process, which are distinct from the initiating phases (Days 1 and 2) and the completing phases (Days 6 and 7). The third Day of an Underworld is usually when the manifestation of the new Underworld, the new level of self-awareness, becomes truly anchored and established, such as in the Cambrian explosion in the Mammalian. The third Day is when the somewhat tentative early phase is left behind. Day 5 is like a breakthrough into the Light, symbolically, which we have three clear examples of in the table. What happened in Day 5 in the Anthropoid Underworld is however not known for certain, but it may possibly have been the emergence of color vision, something that would fit well into the periodic system.

When it comes to comparing the Nights, there is insufficient information to allow comparison in the first half of any of the Underworlds, but when it comes to Night 5, we can see a common pattern of destruction in at least the two lower Underworlds. In one case, the Cellular, Night 5 meant the destructive meteor bombardment of the early Earth, making it temporarily inhospitable to life, but at the same time the return of water and lighter elements, preparing the ground for life to emerge in Day 6. In the Mammalian there was at the beginning of Night 5 (which was ruled by Tezcatlipoca; see tables 2.3 and 2.4) the Permian-Triassic extinction that led to the disappearance of 97 percent of all marine living species,[22] but also to the emergence of the gymnosperm plants.[23] Night 6 in the Cellular Underworld meant the extinction of most anaerobes, but also the emergence of an oxygen atmosphere that paved the way for the highest expression of this Underworld, the aerobic eukaryotes, during Day 7. In the Mammalian Underworld Night 6 meant the emergence of the angiosperms that literally prepared the ground for the higher mammals in Day 7. Hence, while Nights often bring about the destruction of earlier life forms, they also tend to transform the conditions for life in such a way as to prepare for the life forms that are to emerge during the following Day, and sometimes the destruction that they bring helps this very preparation.

As we may see through comparison of the columns of the periodic system, there are significant qualitative commonalities in the progressions through the thirteen Heavens in the different Underworlds, even though the types of phenomena that the Underworlds serve to develop are quite different. We may see this as evidence that each Underworld of the Mayan calendar is a distinct level of evolution driven by a wave movement that in principle is the same in all of them. It is these commonalities that account for the Mayan symbolism of a progression of steps from seed to mature fruit through the thirteen energies, as was shown in table 2.3. We may also see that the various deities seem to match the symbolic qualities of Light and Darkness of the seven Days and six Nights and perhaps come to understand why the specific deities were associated with the time periods that they were. Quetzalcoatl, for instance, who was referred to as the god of light, rules Day 5, which often manifests as forms of breakthrough to light, such as the full transition to land. Tezcatlipoca, the god of darkness was, on the other hand, associated with time periods we might qualify as dark ages, such as the meteor bombardment of the early Earth or the massive Permian-Triassic extinction. In conclusion, the wave movements of evolution generated by the Cosmic Tree of Life in all the Underworlds follow the same progressive sequence symbolized by the thirteen deities.

The common patterns discernible in the periodic system of biological evolution—the seed-to-fruit progressions of the Days and the destruction/ preparation character of the Nights—and the accelerated evolution toward increased self-awareness and intelligence, as well as the generally very good concordance with the modern datings of fossils at cycle shifts, show that the Mayan calendar system is an expression of a creation process of the biological species. Critics may quibble over certain datings or lack of information about certain calendar periods, but the logical and coherent picture of four distinct levels of evolution cannot be shaken by minor changes in any of the dates of emergence. The very fact that an exact time plan for the evolution of life has been verified proves without a shadow of a doubt that Darwinism is a false theory.

It should be added that several ancient cultural and spiritual traditions of our planet, and not only the Maya, were aware of the reality of this basic wave movement of evolution and have expressed it in one form or another. The Jewish-Christian tradition with God's creation in seven Days and six Nights described in Genesis is a case in point, while the Qur'an speaks of six Days.* The latter possibly is a reference to God "resting" on the seventh Day and would then be consistent with the Judeo-Christian view. The Hindus and the Buddhists, moreover, count as holy the number 108, as in the "108 Transformations of Shiva" and in several other symbolic contexts. The number 108 is likely a reference to the entire periodic system of evolution—including all of the nine Underworlds—as there are twelve transformations between Days and Nights in each of these nine Underworlds and so a total of 9 × 12 = 108 transformations. Shiva, as we may recall, was the god of both creation and destruction—destroying in order to create space for the new—and it seems natural to associate him with a wave movement that seems to bring both. Hence it can be stated that on an intuitive, spiritual level, ancient peoples from the whole world were once aware of the basic

---

*Qur'an, Surah 25:59 (Tahrike Tarsile Qur'an). "[Lord] Who created the heavens and the earth and what is between them in six periods and He is firmly established on the throne of authority . . ."

pattern of evolution. Yet it was only the Maya that were able to express these wave movements in terms of exact time periods that are useful for a modern study of evolution. Considering that the intellectual contributions of Native American peoples in general have been neglected by the world community, it is ironic that this people developed the by far most advanced "analytical cosmology" that this planet saw in ancient times.

The clear periodicities that we see in the biological evolution on Earth during the different Underworlds and the common patterns that we have now discovered put the discussion of the causes and mechanisms of biological evolution in an entirely different context, a cosmic context. Because of the sudden shifts between classes of organisms in the different time periods in table 4.5, we may recognize that they are, as argued earlier, quantized, which is the very basis of a periodic system. Thus as it ultimately was the periodic system of the elements that prompted Niels Bohr to recognize the quantized nature of the atom, it is the periodic system of evolution discovered here that prompts us to recognize the quantized nature of the Cosmic Tree of Life, something that will be further discussed in chapter 6. The fact that biological evolution proceeds in leaps on different levels, like "quantum jumps" apparently induced by transformative pulses rather than at random points in time, also allows us to conclude that not only Neo-Darwinism, but also the theories of formative causation, spontaneous self-organization, or the more recently popular epigenetics are, at the very least, insufficient as explanations for biological evolution. The reason is that in none of these proposed theories of biological evolution does the emergence of biological species occur with the kind of predetermined periodicity as shown in the Mayan calendar. Another significant conclusion is that *even though there is a common direction of evolution toward increased self-awareness, this cannot be understood as proceeding along a single line.* Instead biological evolution proceeds on four distinct levels, each based on a new level of self-consciousness of the biological organisms and each developing according to its own rhythm and particular purpose.

CREATED IN THE IMAGE OF THE COSMOS

What then is the purpose of biological evolution? To see how biological evolution has a direction toward a specific purpose, we need to look at the four lowest Underworlds of evolution, with their distinct purposes, separately. The purpose of the first Underworld seems to be to create cellular life. The first Underworld in fact brings the single eukaryotic cells to quite a high degree of specialization so that they are able to move by means of their flagella in a way that seems intentional. These also have a rudimentary ability to respond to light. The purpose of the second Underworld is to develop this initial "consciousness" into symmetrical bilateral animals with heads, fully developed eyes, and brains that are able to process the visually accessed information. The third Underworld primarily develops the erect posture, which amounts to a vertical central nervous system and hands that are free. Very likely, incidentally, the third Underworld affects the evolution of the monkeys and drives them to the erect australopithecines.* The fourth Underworld develops the brain volume of the humans at an especially rapid pace.

In short, the first Underworld introduces cells; the second, multicellular bilateral organisms; the third, vertical organisms; and the fourth, organisms with large brains. Looked upon in this way, we can see that each Underworld has a specific purpose related to the three dimensions of space, and that there is a strong geometric aspect to biological evolution that becomes evident only when you look upon the roles of the different Underworlds separately. Once biological evolution is looked upon as a result of four distinct processes generated by four sequentially activated Underworlds, it becomes evident that biological evolution has a fundamental purpose of developing the three-dimensional anatomy of the various species, and this is not in any sense unplanned. As we saw

---

*It may also have had parallel effects in other groups of mammals, such as felines, dogs, bears, and sea lions (Prothero, *Evolution,* 291), which in the past 41 million years likely have developed a greater intelligence and playfulness.

in chapter 1, the three dimensions of space are fundamentally generated by, and expressions of, the Cosmic Tree of Life, and we can therefore draw the conclusion that *the purpose of biological evolution is to create organisms in the image of the Cosmic Tree of Life.* That biological evolution is fundamentally related to the three dimensions of space will also be important to remember as we elaborate on its actual cell biological mechanism in chapter 7.

Even if all organisms to various degrees reflect the Cosmic Tree of Life, it should be noted that the highest of the four Underworlds of biological evolution, the Human Underworld, leads up to a species, *Homo sapiens,* that more than any other reflects this tree in its design. It is more to the point to talk about the ascent of humankind, rather than of our descent. The purpose of biological evolution is, in other words, not only to create a large and varied biosphere, but also to create a special being in the image of the Cosmic Tree of Life and the cosmos. We may see this not only in that our lateralized brains are reflections of the polarized field of the universe, but also in our erect postures reflecting the Cosmic Tree of Life. Moreover, our brains have an estimated 100 billion brain cells,[24] which seems close to the common estimate of stars in our galaxy, 100–200 billion,* or the estimate for the number of galaxies in the universe, which, based on information supplied by the Hubble telescope, is about 125 billion.† Even though some of these figures are uncertain, they indicate that the numbers of cells in a brain, stars in the galaxy, and galaxies in the cosmos are of the same order of magnitude, and so our brains in principle have been created in the image of the cosmos. This illustrates a second aspect of how human beings have been created in the image of the cosmos.

If we accept the claim made in the previous section that every

---

*Based on the estimated total mass of the galaxy, the number of Sun-like stars is about 175 billion, but based on an average star size, which is smaller, the number would be higher.

†From photographs taken by the Hubble Telescope of the Deep Field it was estimated in 1999 that the number of galaxies in the universe is about 125 billion.

Underworld develops a more marked level of self-consciousness in the biological organisms, then this will have some very important consequences. If the purpose of biological evolution, as has been stated here, is to create an organism in the image of the Cosmic Tree of Life, and if every Underworld plays a distinct role in the development of this three-dimensional anatomy, then this would mean that the self-consciousness of biological organisms is developed in parallel with the creation of their anatomies in the image of the Cosmic Tree of Life. This means that the evolution of self-consciousness is a significant part of the cosmic time plan and that consciousness is not something that is out there diffusely floating around unrelated to the process of biological evolution. This insight needs to form the basis for any true theory about the nature of consciousness. *Consciousness is something that emerges only from the relationship to the Cosmic Tree of Life, and it is through their reflection in the Cosmic Tree of Life that biological organisms may become fully conscious of who they are.* Fundamentally, then, consciousness is based on a relationship, and the evolution of consciousness implies the evolution of this relationship.

If it is the purpose of biological evolution to create a self-aware organism in the image of the cosmos, which then is able to relate to the cosmos, we should not be surprised that with the human being at the end of the fourth Underworld an organism arose, which in principle has the potential to understand the vastness of a universe with 100 billion galaxies. This particular organism has also been created with an erect posture and hands free to manipulate its environment and thus in a sense an ability to create a universe of its own. Biologically speaking the purpose of this creation has been fulfilled with the completion of the four lowest Underworlds, and when it comes to the generation of species with large brain capacities *biological evolution is already over.* Hence, human beings are very special in the cosmic creation scheme. Our species already has the body that it needs, and how it is used is up to us. I feel this conclusion conforms to the intuition that many people have about this. Who seriously thinks that a new, more intelligent species would evolve from us and outcompete us on this planet?

It is for the very reason that human beings, more than any other species, have been created in the image of the cosmos that we are special. We are alone in having been endowed with a brain capacity that allows for a mental and spiritual evolution from which civilizations may be developed. In consequence, once this being, the human being, with an ability to be in resonance with the entire universe, emerged there is no reason for biological evolution to continue (even though minor genetic drifts may still take place among lower animals). The continued evolution of the universe took a different turn as the fifth Underworld began, about one hundred thousand years ago, and evolution started to affect the human mind and what this projects onto the external world. These are however matters outside the scoop of this book.

From this point and onward I will hence consider the Tree of Life hypothesis as a theory and will refer to it as a such, since it is consistent with far more observable data than any alternative theory of evolution. The fact that evolution follows a strict time plan may have been surprising or even shocking to many, considering that for a long time evolution has been claimed to be chaotic, random, and purposeless. In the next chapter, we will go back to the basic physics and chemistry of the universe and there find more evidence along the same line, namely that the generation of life is the primary purpose of this universe, and mathematics will back this up. After that we will return to biological evolution to study its larger context (chapter 6), biochemical basis (chapter 7), and actual mechanism (chapter 8). We will then find that different aspects of biological evolution proceed synchronistically at several different levels of the cosmos. Finally, we will in chapter 9 arrive at the most significant insights and existential consequences of the new theory.

5

The Constants of Nature Are Not Fine-tuned for Life by Accident

THE EVOLUTION OF METALLICITY IN
THE CELLULAR UNDERWORLD

The Tree of Life theory is not just a theory about how biological organisms emerged and evolved on our particular planet. No serious theory about the origin and evolution of life can look upon this as merely a question of "biology" in isolation from the universe at large. This theory is also about the *cosmic context* of life and how this has evolved to produce the necessary physico-chemical conditions for biological organisms to emerge. These two processes, the emergence of life and the emergence of a life-supporting environment, are, as we shall see, synchronized, something we will study from different perspectives. In this section we will focus on how the physical and chemical properties have evolved in the star systems of our galaxy to prepare for the emergence of life on Earth. One such property is the so-called metallicity that a star system must have to a sufficient degree in order for metal-based terrestrial planets and carbon-based life to emerge. It should be said that metallicity is not exactly what it sounds like. Metallicity is a parameter measured by astronomers as an expression of the proportion of elements in a star that are heavier than hydrogen and helium. The concept of metallicity thus deviates from what would be expected from its colloquial meaning. Hence, all atoms with higher mass numbers, including nonmetals such as carbon and oxygen and other biological elements, as well as true metals such as iron and uranium, will increase the metallicity of a star. Metallicity is an indicator not only of the quantity of metals in a star system and in its planets but also an indicator of the extent of existence, at least in very general terms, of the elements from which biological organisms may be formed.

Throughout the Cellular Underworld the metallicity did indeed increase over time in the galaxy to an extent that did allow for the emergence of cellular life (figure 5.1). Only with such a diversification of chemical elements, of which metallicity is a measure, could the biochemical materials for cellular life emerge. Hence metallicity is a parameter of critical importance for the evolution of life. In fact, in a sense, the diversity of atoms with different atomic numbers provides something that may be likened to the letters of the most basic language of the universe, the very foundation of its complexity, without which no diversity of life could ever evolve or even exist in the first place. If there were only one element in the universe, for instance hydrogen, the universe would lack the complexity necessary for the generation of life.

It is believed that the only elements formed in the big bang were hydrogen and helium (with small traces of lithium) and that the first generation of stars (Population III) was formed exclusively from these. Yet through a combination of nuclear reactions in the stars and the emergence of a series of partly overlapping populations of stars with different life spans, the metallicity of the matter in the galaxy came to increase. In other words, in the initial Population III stars, a series

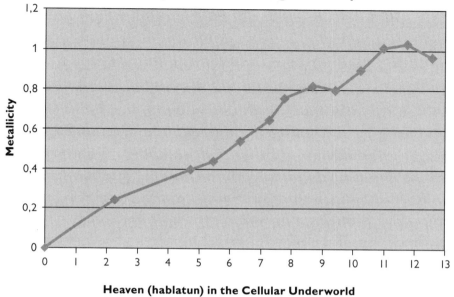

Figure 5.1. Metallicity as a function of age in nearby stars. The diagram shows the metallicity in a sample of stars in the solar neighborhood. Older stars have a lower metallicity because they were formed from an interstellar dust that had a lower metallicity. Toward the end of the Cellular Underworld, there is only a slow increase in the metallicity, which then plateaus. Adapted from Jean Audouze and Guy Israel, eds., The Cambridge Atlas of Astronomy, *3rd ed. (Cambridge: Cambridge University Press, 1994).*

of nuclear fusion reactions, elucidated by Fred Hoyle, took place that produced higher elements, and as a result the metallicity of these stars increased. It is believed that all of these early Population III stars eventually exploded as supernovas in the first few billions of years after the big bang and that this is the reason that no such Population III stars can now be found in our galaxy. The material (with a higher metallicity) released in these supernova explosions was then distributed throughout the rotating galactic disc so that the next generation of stars (Population II), made from the new material, from the very beginning had a metallicity that was higher than that of the earlier generation. After these Population II stars also had exploded as supernovas, the extrastellar dust in the galaxy attained an even higher metallicity. From this was formed the third, and current, generation of stars (Population I), which our own Sun belongs to, which had a yet higher metallicity from the very outset. Only around this Population I generation of stars are there reasons to believe that planets with sufficient metallicity for cellular life could come into existence. Since the formation of the long, complex molecules of life, like proteins or DNA, directly depends on the availability of elements such as carbon, nitrogen, oxygen, and sulfur, it seems likely that they could have been formed only around Population II and Population I stars, and most likely the twenty different elements that make up biological organisms[1] would only exist in sufficient quantities on Population I stars.

The metallicity of a star system also indirectly determines what kind of planets it may harbor. For instance, if the metallicity was too low, an Earth-like planet with sufficient mass in its metallic core to keep oceans and atmospheres on its surface could not form. A life-harboring planet must also have an orbit that is stable to help ensure a suitable temperature for life. It turns out that the stability of the orbits of a planetary system is dependent on its metallicity. If, for instance, the metallicity of a star system is too high, Earth-like planets close to the star will be destroyed by large and massive Jupiter-like planets that move through the orbital region of the Earth-like planets and end up revolving around their stars at a very close distance.[2] A large proportion of the known exoplanets—planets around stars other than our own Sun—in fact seem to be such Jupiter-size planets that due to their high metallicity are likely to have destroyed smaller ones on their course[3] and thus ended up too close to their stars to have a temperature hospitable for life. At any rate, the metallicity of a star system must be high enough for sufficient quantities of higher elements to be present for biological organisms to emerge, but not so high that a planet where they may thrive would be destroyed.

Figure 5.1 shows the metallicity of a few star systems in our solar neighborhood as a function of their age. The older stars, belonging to older generations, have a lower metallicity; there was a lower proportion of higher elements in the extrastellar dust at the time they were formed.[4] This diagram shows that during the course of the Cellular Underworld the metallicity of stars increased, and toward the end of the Underworld it plateaued at a level that we know is conducive to the formation of Earth-like planets that may harbor life. We know this because indeed we live on one of them.

It should be noted as an important aspect of galactic evolution that metallicity does develop, that it is conducive to life in at least some of its star systems in the habitable zone of the galaxy, and that it cannot be too high or too low if this is to happen. Even though we may currently be unable to say exactly how narrow the range of metallicity must be for a star system to create life, it is easy to see that this parameter is directly related to the age of the star system (see figure 5.1), meaning that life could emerge only within a certain "window of time." This window of time probably commences around the beginning of Day 5 in the Cellular Underworld, since the average age of Population I stars harboring Earth-like planets has been estimated to 6.4 Gyr,[5] although our own Sun is usually estimated to be a little younger (about 5 Gyr) and the Earth to have solidified about 4.6 Gyr ago. These age estimates are however based on measurements of solid materials,[6] and the process of formation of the solar system may thus have started somewhat earlier, already in Day 5.

The increase in metallicity in the thirteen Heavens of the Cellular Underworld is broadly parallel to the increase in the number of cell types in the thirteen Heavens of the Mammalian Underworld (figure 4.4), although the former spans a time period that is more than twenty times longer. Together with the increase in brain volume in the Human Underworld (figure 4.5), these curves show that the different Underworlds play distinct roles in the total scheme of biological evolution and serve to increase the complexity of the universe in different ways. What is noteworthy about these curves is that not only do they show a complexity increase from seed to mature fruit in the different Underworlds, but they also plateau in a way that provides an optimal basis for the next level of evolution.

Because the metallicity increase in the Cellular Underworld is a result of a galactic process, it seems there must also be a direct relationship between galactic evolution and biological evolution, and to account for this a nested hierarchy of Halos was outlined in chapter 1. Without the circulation of stars through the extrastellar dust of a galactic disc, the processes described above leading to an increase in metallicity would have required enormous amounts of time to manifest, or perhaps (and more likely), would not have happened at all. Through its rotation around its axis, the Milky Way plays a critical role in making chemical elements available for the emergence of life throughout the galaxy. This role we may liken to that of planet Earth, which makes a number of elements available to its organisms through the atmospheric and oceanic circulation created by its rotation around its axis.

From this, we can clearly see that the emergence and evolution of life takes place within, and directly depends on, different levels of organizations of life, the Earth and the Milky Way, as well as on the integration of the former in the latter. Life simply would not emerge in a universe that is not organized in a nested hierarchy of vortices (figure 1.5), since the rates of rotation of the axes, the Trees of Life on the galactic and planetary levels, are critical factors for the emergence of life. Thus, the physico-chemical conditions for life did not just accidentally "pop up" at our particular planet. They emerged because the matter in the universe accreted into the different populations of stars at certain rates and because the higher elements were formed from lower at a certain rate in these stars. Later these elements were made available for new star systems and the metallicities of these depend on the rate of rotation of the galaxy. Given these conditions, the existence of a planet like our own that has the right composition of elements not only for biological life to emerge but also for its organisms later to be able to build cars and computers requiring many special metals, might not be such a remarkable accident as it would at first appear. It may exist because the forces of nature and their constants at the galactic level are fine-tuned for this to happen. Before we discuss this, however, we need to take a look at the origin of our galaxy.

GALAXIES AND THE MYSTERY OF "DARK MATTER"

The Tree of Life theory postulates that the universe is a nested hierarchy of different semiautonomous organizations of life that are synchronized to support the creation of the conditions for life on Earth and on similar planets in other solar systems. In the previous section we saw that it is the role of the galaxy to provide the right metallicity for star systems to form where biological organisms may thrive, and a galaxy may really be looked upon as a womb for planets with life. In order for it to be able to exert this role, its star systems need to behave in accordance with its mechanical laws, but it also needs to function as a system that is partially autonomous. The attainment of the autonomy necessary for the generation of life is the topic of this section.

It is today believed that galaxies start out as small vortices, which increase in size by mergers with other such vortices, and over the first billions of years of the universe these grew to large spiral and bar galaxies. Elliptical galaxies, which sometimes are much larger than spiral galaxies, seem to have emerged as two or more spiral galaxies collided at an unfortunate angle so that the ordered movements of their star systems were lost. Our own Milky Way is now believed to be a bar galaxy with two arms (see figure 5.2).[7] Our solar system is located about two-thirds of the distance between the center and the periphery.

It is not understood why the stars stay together in the organized structures of the galaxies, since their total masses are not, according to Newtonian mechanics, large enough to exert the attractive power

Figure 5.2. The Milky Way galaxy and the position of our solar system. Simulated picture of our own Milky Way galaxy as viewed from a point above it. Our solar system is located about two-thirds of the distance from the center to the periphery in what is considered a habitable zone of this bar galaxy. The solar system completes a revolution around the galactic center about every 250 million years. From R. Hurt, GLIMPSE Team, "Astronomy Picture of the Day," http://antwrp.gsfc.nasa.gov/apod/ap050825.html.

necessary for this. Something else needs to be defining the galaxies as systems and maintaining their coherence. Another thing that Newtonian mechanics, which near-perfectly describes movements on the level of the solar system, cannot explain is why the speed of revolving stars in a galaxy does not decrease more at higher distances from the galactic center. While the rates of revolution of planets in our solar system decrease according to Kepler's third law* (as a function of the square root of the cube of their distances from the sun, which is a special case of Newtonian mechanics), the corresponding rates of stars in the galactic context are, in contrast, roughly independent of their distances from the center. There are also several other aspects of the movements of galaxies that Newtonian mechanics cannot explain, anomalies that cannot be dismissed as minor, but indeed are very large.

Physicists have been grappling with these inconsistencies of galactic mechanics for a long time, and the explanatory theory that has received the most attention is that the universe is filled with an invisible "dark matter." Thus if such "dark matter" is distributed in a certain way, it would purportedly explain most of the anomalies. In the planar disc region of the Milky Way, where our own Sun has its orbit around the galactic center, the amount of "dark matter" would need to be ten times as large as the visible matter in order to explain why the Sun stays in its approximately 250-million-year-long orbit around the center of the galaxy. In the large-scale perspective of the universe as a whole, the percentages of "dark matter" and visible matter would need to be 20 and 5, respectively, to explain its mechanics[8] (75% of the "mass" of the universe is now believed to be contributed by a so-called dark energy).

---

*Kepler's third law states that the squares of the times of revolution, t_i, of the planets are proportional to the cubes of their mean distances, a_i, to the Sun:

$$\frac{t_1^2}{t_2^2} = \frac{a_1^3}{a_2^3}$$

Kepler's original law was however only approximately correct and was corrected by Newton as follows:

$$\frac{t_1^2}{t_2^2} = \frac{a_1^3(M + m_1)}{a_2^3(M + m_2)}$$

where t_1 and t_2 are the times of revolution of two planets, a_1 and a_2 their distances from the Sun, m_1 and m_2 their masses, and M the mass of the Sun. Kepler's third law represents such a good approximation of the observable movements of the planets because the masses of the planets are very small in comparison to that of the Sun.

The hypothesis is that there is some kind of "dark matter" that has been around for quite some time. Astrophysicists have done much to find out what it may consist of. Even if several possibilities, such as brown dwarf stars, have been suggested, no positive explanation has however been widely accepted. "Dark matter" seems to behave in a way that is not really consistent with any accreted form of matter or known elementary particle, and there is no direct evidence that any such "dark matter" actually exists. All that is known is that "something" makes the visible matter of galaxies behave in ways that are not expected from Newtonian (or Einsteinian) mechanics. Yet this invisible "something" has been assumed to pervade the universe, and visible matter is thought to be attracted to it and galaxies in fact be organized by it. At least in the Milky Way galaxy, this "something" is assumed to have a spherical distribution, with its center right at the center of the galaxy's visible mass. Recently an interesting observation was reported, where even if "dark matter" was not directly observed, its supposed effects were seen to distort objects behind it similarly to a spherical lens.[9] This finding lends direct support to the idea that "dark matter" is spherically distributed.

Since this spherical "something" apparently provides a stronger attractive power than the visible mass of a galaxy, it is believed that "it" in fact has attracted the visible matter rather than the other way around. Thus it must have been instrumental for the organization of the galaxies, the accretion of matter in them, and the mechanics that they display. "It" has also been the chief factor for creating the shapes of the galaxies as well as determining the speeds of stars revolving around their centers. Because of this crucial organizing role, it seems a good idea to look at this "something" in light of the commonsense anthropic principle, which subordinates the physics of the universe to its purpose of generating intelligent life. We should then ask what effect this "something" might have had on the emergence of life in the galaxy. The answer from this perspective is that the "something" does at least two things. "It" serves to keep stars together in galaxies, and "it" keeps them revolving in a disc around their centers, and both of these things are necessary for the possible emergence of life in their planetary systems. As I pointed out earlier, if the extrastellar dust of high metallicity in the galaxy, produced by sequential generations of stars, had not been circulated in the galaxy so that it was made available to new stars-to-be, then life would not be possible in planetary systems. In order for new stars with a sufficient metallicity to emerge, within the necessary time frame of the Cellular Underworld, stars need to revolve in the galaxy at a sufficient speed. Their speed of revolution, and the fact that the stars stay in the galaxy to begin with, then directly depend on the existence of this "something" that modifies galactic mechanics. If this "something" did not exist, the stars would leave the galaxies or not attain the metallicity needed for life. Hence, it seems as if the emergence of life anywhere in our galaxy critically depends on this "something," and that "it" in fact is a life-generating phenomenon. In the absence of "it," only super-dense galaxies would be able to hold the matter of the star systems together, but because of the high density of matter in such hypothetical galaxies, life there would be imperiled in any of its star systems by the disturbing effects of neighboring stars on its orbits, not to mention by the increased presence of nearby supernova explosions or radiating neutron stars. It consequently seems that the emergence of life hinges on "something" that modifies Newtonian mechanics in such a way that a galaxy keeps together at a relatively low density of visible matter.

Life could emerge only in a galaxy that has a sufficiently high density for matter to accrete, but where at the same time this density is not so high that life-harboring planets are extinguished by too many interactions with neighboring stars. Thus galaxies could easily have miscarriages, and it seems that we have again demonstrated a case of fine-tuning in nature, which in this case is affected by the organizing power of "something" that is just right for the generation of life. The reader has probably already realized that this "something," spherically organized around the spinning axis of a galaxy and serving to hold it together as a structure, is nothing other than what I initially referred to as a Halo, in this case the invisible galactic Halo. As all Halos, it serves to create an organization of life at a certain level and to provide a certain autonomy, and hence, as we have now seen, only in a context determined by a Halo could life emerge in a galaxy. Since the effects of these Halos appear to be considerably more powerful than the visible matter that they organize, it is logical to conclude that they play the primary role for the origin of galaxies.

I suspect that no physical "dark matter" will ever be found, but even if some particle given an exotic name was brought forth, it would still only be another name for the life-organizing galactic Halo. This discussion of "dark matter" also shows to what extent the terminology used by physicists predisposes people to embrace a mindset in which the universe has no purpose. Scientific terminology is not neutral, but implies a particular metaphysical worldview, and when scientists call the Cosmic Tree of Life and its Halos the "Axis of Evil" and "dark matter," this biases not only their own thinking, but that of the informed public as well. In reality, a galaxy, such as our own Milky Way, is a womb for planetary life, in the sense that all of the life forms of this planet—its plants and animals, as well as we human beings—have been generated from material in a nested hierarchy of life defined by tiers of Halos. Without a Halographic context there would not have appeared any new generations of stars with a high enough metallicity to produce life within the time frame of the Cellular Underworld.

A NEW BASIC CONSTANT OF NATURE DEFINING THE RHYTHMS OF EVOLUTION

In the previous chapter we saw that the timing of a number of biological emergences adhered very closely to the Mayan calendar shift points, which shows that these are determined by a basic rate of evolution in the universe that is reflected in this calendar. An interesting thing to study is how these points of emergence relate to the increased metallicity in the galaxy and if the basic constants of the universe relate to this predefined evolutionary rate. In chapter 1 it was pointed out how the evolution of metallicity of the entire universe depended on the fine-tuning of two of the basic constants of the universe, α_s and α_w, expressing the strengths of its strong and weak forces, respectively. As mentioned earlier, if α_s had been lower, no complex life molecules, such as those based on carbon chains, could have emerged, and, on the other hand, if α_s had been higher, the carbon would have been burnt to oxygen and other higher "metals" so fast that life could not have formed in this scenario either. These constants need to have values within a narrow range in order to produce the right metallicity for life to emerge. This has been pointed out many times before, but what we will discuss here is the evolutionary context for this fine-tuning.

What then needs to be added for our understanding of the fine-tuning is that for the emergence of life, it is also critical *when* the metallicity is right. The metallicity, and hence the constants α_s and α_w that it directly depends on, needs to be fine-tuned to the rate of evolution so that it has the right value in a certain "window of time" that is conducive to the emergence of life. Unless the right metallicity is present in the "window of time" of Days 6 and 7, the prokaryotic and eukaryotic cells, respectively, would not have emerged. If, for instance, the wavelength (period) of cellular evolution was half of what we see in the Cellular Underworld (e.g., 1/2 hablatun = 0.63 Gyr), the cellular Halos would have attempted to create eukaryotic cells about seven billion years ago, at a time when the metallicity would not have been sufficient for such cells to form. Or, if the rate of evolution had been twice its actual rate, the Halos would have attempted to create cells maybe 14 billion years into the future, at a time when our solar system is no longer expected to exist. It is difficult to assess exactly how narrow the range for the constant rate of evolution must be for the emergence of life, but in our own solar system, which seems to have a suitable metallicity for this, the hablatun duration could not deviate by more than 30 percent from its actual value—and probably much less.

The rate of evolution must be a fundamental parameter of the space-time organization generated by the Cosmic Tree of Life in order for this to fulfill its purpose. It seems however that it would be more relevant to reverse the example and look upon the constants a_s and a_w as being constrained by the hablatun rhythm in order for cellular life to emerge in the predetermined Days. As mentioned in chapter 1, the rate of accretion of matter in the galaxy also critically depends on the ratio α_G/α_{EM} between the strengths of the forces of gravitation and electromagnetism. (Halos also must modify Newtonian mechanics on the galactic level.) This means that for a galaxy to be able to generate life in any of its star systems, the strengths of all four forces of nature, α_G, α_{EM}, α_s, and α_w, need to be fine-tuned to the evolutionary rhythm of the Cellular Underworld, given by the hablatun. Hence, we now for the first time have a context for, and a background to, the fine-tuning of some of the constants of nature. This is especially true if, as mentioned earlier, the original rate of expansion of the universe in the big bang was determined by the Cosmic Tree of Life.

If we accept that the evolutionary processes of life emanate from the Cosmic Tree of Life, and that they adhere to the tun-based system of the Mayan calendar, we no longer have to marvel at how fine-tuned the various constants of nature must be for the emergence of life. This simply becomes a logical consequence of the fact that these constants were determined in synchrony with the Cosmic Tree of Life as this emerged in the big bang. The strengths of the four forces of nature, and several other constants, are simply subordinated to, and locked into, the particular values that they have by the rate of evolution and the rate of the original expansion of the spatial dimensions of this universe. There are the fundamental parameters of the space-time organization of a life-generating universe defined by the Cosmic Tree of Life.

Obviously, no one can be forced into accepting this new big bang scenario, where the emergence of the Cosmic Tree of Life is the determining event. Yet it should now have become clear that it provides a logical explanation for the fact that the constants of nature are fine-tuned for the emergence of life. Things become much simpler with the commonsense anthropic principle: A universe that does not have an evolutionary rate defined by the tun system is not able to generate life, and conversely, in a universe where the evolutionary rate is defined by the tun system, life may be generated, provided that all the other constants of nature are constrained by this tun system, and that there exists a nested hierarchy of Halos delimiting the various systems.

The rate of metallicity increase in the Cellular Underworld thus has to be fine-tuned to its hablatun rate of evolution in order for life to emerge. Since the rates of evolution of all the other Underworlds are generated by dividing the hablatun rate by different multiples of twenty (table 2.1), it seems that it would be sufficient to include only one of these rates in a theory about the evolution of a life-generating universe. It would then seem natural to use the tun as the basic constant of nature expressing the rate of evolution, since the other evolutionary rates are all defined by this.[10] It may be noted that the tun constant, in addition to being an expression of creation by the World Tree in Mayan cosmology, is also considered to be the prophetic year (of 360 days) of the Bible,[11] where its relevance is also directly implied in the Book of Revelation.* As a number, 360 incidentally also has interesting properties and is the number below 500 that can be factorized in the most possible ways.

There is however another very consequential conclusion to be drawn from the previous discussion. Since it is the Cosmic Tree of Life that is sending out creative pulses with a hablatun rhythm, we really have no reason to believe that our planet is the only one harboring life in the cosmos. It would instead seem that many planets in many galaxies, meeting certain conditions to be discussed in the next section, must have the potential to develop life *in synchrony with*

---

*The 1,260 days described in Revelation 12:6 are identical with the "time, times and half a time spent in the desert" (Revelation 12:14). The Official Swedish Bible translation of 1981 is however more explicit than the New International Version and reads: "a time and two times and half a time" (Revelation 12:14). This means that 1,260 days equal 3.5 "times," or in other words, that 1 "time" equals 1 tun, which equals 360 days.

the process that has developed on our own planet. This is because life as such ultimately does not originate on Earth or in the galaxy, but from the Cosmic Tree of Life. Since the constants that define the universe are the same everywhere, biological organisms in principle would evolve at the same rate everywhere. In our particular case it is the Milky Way and the Earth that have provided the environments and building materials for life to emerge in accordance with the cosmic time plan.

THE FINE-TUNING OF THE FORCES OF PHYSICS TO THE ORBIT OF THE EARTH

The Earth's orbit is the measure of all things.

JOHANNES KEPLER

Not very surprisingly from this background the physico-chemical conditions on our own particular planet in numerous ways also seem fine-tuned to, and more or less optimal for, the emergence of life. Earth has an atmosphere with a concentration of carbon dioxide that is high enough for the existence of carbon-based life, but not so high that it creates a runaway greenhouse effect. Earth has substantial amounts of liquid water, but in this water there are also continents of solid matter with varied climates that provide for a variety of different ecosystems where higher forms of life, and later also technological civilizations, may emerge.

At closer inspection we may realize that many, if not most, of the conditions that make the Earth a hospitable planet are consequences of its distance to the Sun, which gives it a temperature that is suitable for higher life. It has been estimated that if the Earth's distance from the Sun was only 5 percent shorter, it would have become a choking greenhouse, and if it were 1 percent farther away, it would have suffered eternal glaciation.[12] Why then is it that the Earth is able to do this delicate rope walk? Why is it located at exactly the right distance from the

sun for life to thrive? This is again related to the tun as a basic constant
of nature inherent in the space-time organization of the Cosmic Tree of
Life. We may understand the emergence of optimal conditions for life
on Earth in terms of Kepler's third law. According to this law, the distance of a planet from the Sun, and thus by implication its habitability,
depends on its period of revolution around the Sun. For the Earth, this
period of revolution is 365.2422 days. This period is close to the tun
(and it is very noteworthy that the year slowly approaches the value of
360 days),[13] a constant that, as we saw earlier on the galactic level, the
constants of the forces of nature have been fine-tuned to. On the level
of the solar system, we are able to see the effects of this fine-tuning in
a new and more precise way because of the immediate sensitivity of the
habitability of our planet to the strengths of these forces.

On the level of the solar system, it is α_s and α_w, the constants of the
strong and weak interactions, respectively, that determine the rates of
the nuclear reactions in our Sun, which in turn determine what the
temperature at different distances from the Sun will be. At a given
strength of the force of gravitation, α_G, these distances from the Sun
are determined by the rates of revolution of the planets according to
Kepler's third law. In order for a planet with a tun orbit to have a certain temperature, these constants need to have specific relationships to
one another. Or, in other words, as a group they must be constrained
by the tun orbit, which is a primary characteristic of the Cosmic Tree
of Life. Yet, for life to emerge on a planet with a tun orbit, there is
also the fourth force of nature, the electromagnetic force, that needs to
have exactly the right strength. The reason is that this force determines
the strength with which matter is held together, for instance when
negatively charged electrons form clouds around the positively charged
nuclei of atoms or molecules. If its constant, α_{EM}, did not have the right
value in relation to the three other forces, α_G, α_s, and α_w, then it would
detrimentally, and profoundly, influence the chemistry, and the ability
of biomolecules to form and be stable, on a planet with a tun orbit. The
constant, α_{EM}, expressing this force, has a value that allows the carbon

atoms necessary for life to hold together and exist in the first place. As we can see, *all four forces of nature are constrained by the tun constant and the commonsense anthropic principle.*

An important consequence of this is that because of the way the constants of nature are constrained by the Cosmic Tree of Life, they do not allow for life to emerge at just any location in the solar system. Instead they are constrained so that all the conditions are optimal for life to emerge on a planet with a tun orbit, since there, on a local level, its polar axis reproduces the Cosmic Tree of Life. In orbits other than the tun orbit, the conditions for biological evolution to higher levels will not be at hand. Earth is hospitable to life because the constants of the four forces of nature, as well as the tun constant, were defined by the space-time organization of the Cosmic Tree of Life at the big bang. Earth also has an orbit whose rate of revolution is defined by this tun constant. This shows that the purpose of the universe to generate life existed from its very beginning with a focus on manifesting this exactly on planets with a tun orbit. Our own particular Earth was "tagged" for life because of its tun orbit, and life has not appeared here because of a series of innumerable randomly occurring events. This reasoning however does not immediately tell us exactly how the strengths of the four forces of nature are set at the values conducive to life, and we need to ask how exactly they are defined by the space-time organization of the Cosmic Tree of Life. To explain this, we will here focus primarily on electromagnetic force.

The strength of the electromagnetic interaction, α_{EM}, which also goes by the name of the *fine structure constant*, has a value that is close to 1/137. For a long time this value has appeared as mysterious to physicists, who have sometimes sought a numerological explanation as to why it has the particular value that it does. Its value has seemed so basic to why there is atomic matter in the first place that it has been speculated that it must be derived from one of the fundamental constants of mathematics. Wolfgang Pauli, the father of the Pauli principle in quantum chemistry, believed that this constant was the most important in all of physics and that the future of quantum theory depended on an understanding of what lay behind it. It is here also relevant that for a long period of his life Pauli was engaged in discussions with Carl Jung to find the underlying forces in the universe generating synchronicities.[14] As it turned out, Pauli died in a hospital room with the number 137, and before he passed away he expressed the wish that he had had a chance to discuss this important synchronicity with Jung. As another example, Richard Feynman, the father of quantum electrodynamics, referred to the fine structure constant as "one of the greatest damn mysteries of physics: a magic number that comes to us with no understanding by man. You might say the 'hand of God' wrote that number."[15] He speculated that it might be related to pi or *e,* the base of the natural logarithms. Although this type of "numerology" is not really *comme il faut* among physicists (since it would easily lead in the direction of revealing a higher meaningful plan), certain scientific notabilities have sensed that there must be some very simple mathematical reason that the constant of electromagnetic interaction, arguably the most important constant in nature that defines atomic formation, has exactly the value that it does. This constant, which is a plain dimensionless number, determines the structure of atoms, molecules, and solid materials together with the behavior of light, whose gross physical properties in principle can be determined as functions of this constant (and another quantum relationship, the ratio between the masses of the electron and the proton, which we will leave aside).

While twentieth-century physicists were not able to identify any convincing mathematical constants underlying the fine structure, partly because such thinking has normally not been encouraged, a revolutionary suggestion was recently made by the Czech physicist Raji Heyrovska, who deduced that the fine structure constant, experimentally determined as 1/137.036, really is defined by the ratio $\phi^2/360$, which equals 1/137.508.[16] From this, we can then identify 360 as the ratio between the tun and the kin, the two periods that play the central role in defining Mayan calendrics, and ϕ, or phi, the so-called golden ratio. The golden ratio is also known as the golden mean and the divine proportion, and it is a mathematical constant that we will encounter several times as we continue to explore the Tree of Life, especially in the last chapter of the book. This number represents a geometric relationship that is recognized as a significant part of sacred geometry and has been widely discussed on the Internet[17] and in several books.[18] In the Tree of Life theory, *phi (φ) expresses the basic organization of space in the universe.*

Before we continue our discussion of the fine-tuning of the constants of nature to the orbit of the Earth, it is here timely to further introduce the golden ratio. This proportion is usually expressed by the Greek letter phi (φ). It was for the first time defined in words by Euclid, who in *Elements* divided a line AC into two sections by a point B such that the ratio of the length of AC to the larger part AB was the same as the ratio of the larger part AB to the smaller part BC. In short, φ = AC/AB = AB/BC (see figure 5.3).

From this geometric definition it may be derived that 1/phi = phi − 1, from which the value of phi may be calculated as $\frac{1+\sqrt{5}}{2}$, which is an irrational number with the value 1.618. . . . What we should especially notice here is that this constant, which by several authors has been considered as underlying the creation of the universe, defines a harmonious relationship between a whole and its parts. Such a suggestion is consistent with our theory, since if this constant defines the spatial organization of the Cosmic Tree of Life, then it would account for an optimal integration of the parts in the whole at the many levels of the nested hierarchy of systems determined by the various lesser Trees

A   B   C

Figure 5.3. The golden ratio. In the golden ratio, the ratio between the whole, AC, and the larger part, AB, is the same as the ratio between the larger part, AB, and the smaller, BC. In other words, AB/BC = AC/AB = φ.

of Life (figure 1.5). For this reason we should not be surprised that this constant may be identified at many different levels in nature and in all life. Because of its widespread manifestations, the real problem for developing a logically coherent theory is in restricting the focus only to manifestations that are relevant for this discussion.

In my view, it is in such a context that Dr. Heyrovska's finding that the constant determining the strength of electromagnetic interactions throughout the universe, $\alpha_{EM} = \phi^2 \times kin/tun = \phi^2/360$, needs to be understood. We here find, in other words, without resorting to complicated mathematics, that *the value of the fine structure constant, α_{EM}, is a direct consequence of the space-time organization of the Cosmic Tree of Life, which is characterized by the constants tun and phi (φ).* It may also be argued that the emission and absorption of light from atoms, including when light is emanating from ourselves, is defined by this space-time organization of the tun and the golden ratio (because α_{EM}, and hence ϕ, is part of the Rydberg formula determining the wavelengths for emission of light). Thus there seems to be some truth to the idea that light has a divine origin in a more literal sense than many would probably think. From these considerations we may realize that once we give up the cosmological principle upon which relativity theory relies, and accept the Cosmic Tree of Life as the primary space-time organizer of the universe, then physics becomes more simplistic and also more meaningful. It also makes sense from a spiritual perspective in that the strengths of the forces of this universe, and as a consequence the generation of life, follow directly from its space-time organization, which is defined by a universal creative source. This is of course exactly what we argued earlier—that the purpose of this universe is to generate life— but we have now substantiated this with the mathematical proof that *life emanates as a direct consequence of the space-time organization of the Cosmic Tree of Life.*

I feel the profundity of these results should not escape anyone, because we may now really in a very direct and dramatic way see that our universe and its constants are designed for the creation of life and that this emanates directly from its space-time organization. For a complete theory, we would still need to work out exactly how the three remaining constants, α_G, α_w and α_s, of the forces of nature are mathematically related to the space-time organization of the Cosmic Tree of Life. Yet by comparison this seems to be a problem of minor importance.

We have found that the values of the constants of nature have not been fine-tuned for life by accident, but that these values are constrained by and logically follow from the fundamental space-time organization of the Cosmic Tree of Life. This reasoning starkly highlights how decisive the choice of including, or not including, the tun and phi constants is for our perception of the nature of the universe. For those who include them, they become the basis of a new physics in which the design plan and purpose of the universe to create life is a logical conclusion. For those who do not include them, it may be possible to continue to present life as an unplanned accident hinging on a number of constants that just "happened" to be right. This is what some cosmologists have done to explain the emergence of life in our particular universe. They have even resorted to suggesting that we are living in a "multiverse," in which a zillion different universes with different constants have been generated at random. They claim that we just happen to live in the one that is favorable to life. I should here say that I am not adverse to the idea of a multiverse as such. Yet in this context such speculations testify to how complicated modern science has become, and how far it has strayed from actual empirical observations, in its ambition to uphold the dogma of nondesign. It also becomes clear that the randomism of modern science is a reflection of its predominant antitheistic ideology. While anyone is entitled to such an ideology, or what we could call an atheist religion, I see no reason that this particular ideology should be allowed to monopolize science and present itself as the only possible truth when in fact it does not foster clarity or simplicity.

Another aspect of the Tree of Life theory is that the rate of biological evolution on our own particular planet according to Kepler's third law is directly related to its rate of revolution around the Sun. We are now presenting decisive scientific evidence that the emergence of life on our particular planet is a direct consequence of the fine-tuning of the forces of nature to the tun orbit of the Earth—a fine-tuning that was initiated already at the big bang—and this points to possible connections between the astronomical cycles of the planets and the evolution of life in the solar system. The connection between its tun-orbit and the emergence of life on our planet is in fact a significant vindication of the work of Johannes Kepler, who worked as an astronomer at the court of Rudolf II in Prague in the early seventeenth century, where he incidentally also did significant work on the golden ratio. Kepler was the first to formulate mathematical laws of astronomy—which may also be seen as the first mathematical laws of nature—and for this particular contribution he has been widely recognized. What has mostly been downplayed however is his basic intuition that the orbits of the planets were defined so as to create a divine celestial harmony. Maybe it is only in our own era that we can begin to appreciate the deeper truth of his scientific vision. Very likely, the synchronicity would have intrigued Pauli and Jung—that it was in Kepler's city, four hundred years after his death, that the golden ratio was found to play a significant role for the emergence of life on our planet, a role that it could play only because of its relationship to the orbit of the Earth.

Finally, we can now also see that the determinism that became evident as we studied the time-wise evolution of the different biological organisms in fact is paralleled by an equally strong determinism in the evolution of the physico-chemical conditions necessary for life. On closer examination it is easy to realize that this could not be otherwise, since in evolution timing and synchronization are everything. A universe designed for generating life cannot allow for randomness either in its biological or physico-chemical parameters, and especially not in the relative rates of evolution of these. For the universe to be able to generate life, the constants of nature need to have relationships to one another within very narrow ranges, which ultimately are subordinated to the space-time organization provided by the Cosmic Tree of Life. We can be certain that many other constants than those already mentioned are also constrained by the space-time organization of the Cosmic Tree of Life. We are here only beginning to formulate a theory of everything about the physical reality in mathematical terms. What is important to note for our purposes however is that by definition such a theory must be a description of a universe with life and not one of dead matter. This incidentally is also the only kind of universe that anyone has ever known to exist.

AN END TO THE FUZZINESS
OF QUANTUM PHYSICS

From the previous discussion I feel it is becoming increasingly clear that the universe is not the result of a big bang, but rather of a big thought (although it is always possible to choose to focus on aspects of reality where it looks random or pointless, as we well know). If the universe is a Big Thought, it is natural to ask: *What are the letters in which this thought is expressed?* Written English uses twenty-six letters, from which all words and sentences are formed, allowing for a potentially enormous variation of thoughts to be expressed. In mathematics these letters would be called discrete, as they do not mix with one another; an *a* is an *a* and a *b* is a *b,* and no single letter can be formed by mixing the two in different proportions. This provides a meaningful analogy for the letters of the universe, the atoms, which are also always discrete; an atom may for instance be an oxygen or a nitrogen, but no single atom is a mixture of two different elements. There are some ninety-two naturally occurring "letters"—chemical elements—in the alphabet of the matter of the universe, and their discreteness is maintained by the phenomenon of quantization, the basis of quantum physics and quantum chemistry.

For life to emerge in a universe it in fact seems as if atomic matter needs to be quantized. As pointed out as we discussed the metallicity of the universe, these atoms also need to exist in fairly exactly defined proportions for life to emerge. No planet hospitable to life could for instance be formed only by the elements hydrogen and uranium in equal proportions. To understand the emergence of matter from which life could be formed, we must study quantum theory at its most basic atomic level. Quantum theory describes atomic structures and the nature of subatomic particles as well as light. It has however been perceived as describing phenomena for which special rules and laws apply that often are in conflict with our everyday experience. Yet the rules of quantum physics as they apply to the atomic level are as much a necessary part of the life-generating universe as the law of gravitation. Without them there would be no letters in the alphabet of the universe through which the Big Thought could be expressed. Despite their traditional reputation as the harbinger of chance and indeterminism, quantized phenomena are really the basis for the fidelity and large-scale stability of nature.

What we will propose here is (of course) that quantum physics is best understood as an integral part of the Tree of Life theory. We have already hinted that there is an Atomic Tree of Life. Moreover, we have said that on a larger scale what makes the Mayan calendar unique is that it describes a quantized aspect of time, something will be discussed in the next chapter. The clearly directed view of evolution in light of this calendar however strongly deviates from the indeterminist and probabilistic view that has mostly been associated with quantum physics, and so this difference in viewpoints needs to be discussed here. Given that some of the constants found to be fine-tuned for the evolution of life play a critical role in quantum physics, we have reason to question whether this theory should fundamentally be considered determinist.

But before presenting a new interpretation of quantum theory, it seems useful to provide a brief historical background on quantum theory and highlight some of its most important tenets. In classical physics, which reigned until the end of the nineteenth century, the laws developed to describe gravitation and electromagnetism were based on the idea that the different parameters were continuous. In 1900 the German physicist Max Planck was however forced to conclude that energy radiating from heated bodies was emitted in the form of quanta, discrete packages of energy, which could not just take any possible value. This was at the time a quite revolutionary conclusion, drawn reluctantly, as it seemed to conflict with everyday experience, in which the temperature of a body underwent continuous change. Nevertheless, based on this concept of quanta, Albert Einstein in 1905 proposed that light not only had an established wave aspect, but also a particle aspect manifesting in the form of photons with discrete energies. With this a mysterious aspect of quantum physics was introduced, the wave/particle duality of subatomic particles, that later was hypothesized by de Broglie to apply to macroscopic objects as well.

In 1913 the Danish physicist Niels Bohr had his stroke of genius and proposed an atomic model, where electrons circulated around the atomic nucleus at defined discrete distances with quantized energies in electron shells. In his atomic model, the electrons circulated around the atomic nucleus much like planets circulate around the Sun. There was however a significant difference in that in Bohr's model, the electrons could circulate only at certain discrete distances from the nucleus, which corresponded to different energy levels. This quantization of the electron orbits explained why these particles were not drawn into the positively charged nuclei of the atoms, something that would have led all the atoms of the universe to collapse. Moreover, Bohr's atomic model successfully explained how quanta of light would be absorbed or emitted when the electrons jumped (so-called quantum jumps) between the different quantized levels and how the wavelengths of these quanta of light were determined by the energy levels of the orbits. Although his model, as we will see shortly, is now considered an oversimplification, it was a successful development in that quantum theory could now logically explain the periodic system of the chemical elements based on the numbers of electrons in the atom shells.

From 1925 to 1928 quantum theory may have had its most significant theoretical breakthroughs. First, Bohr's German student Werner Heisenberg was able to provide a mathematical formulation of the theory that included the famous uncertainty principle, which proved that the position and momentum of a subatomic particle could not both be measured with certainty. This contributed much to the perception that quantum theory, in Einstein's words, was "fuzzy." Second, the Schrödinger wave function was proposed as a mathematical tool to describe the mechanics of electrons at different energy levels in atoms. From this model, the probability of finding an electron in a certain location around an atom could be calculated as the electrons were distributed in orbitals. These are better described as electron clouds, which replaced Bohr's view of electrons only moving in certain orbits. This made matter seem even more "fuzzy," and reality was by some reductionist popularizers presented as probabilistic, since it was only the probability of finding an electron in a given location that could be calculated from the Schrödinger equation.

As we can see in figure 5.4, electron orbitals have different geometric shapes, which correspond to different waves and can be determined from the Schrödinger equation. There can never be half an electron wave in an atom, only a full crest and trough. A very important consequence of this is that the electron orbitals always correspond to certain quantized states, meaning that the chemical elements are discrete and distinct. The electrons in these orbitals are strictly defined by a set of quantum numbers that are integers, and new and higher atoms are formed as electrons with higher quantum numbers are added. According to the Pauli exclusion principle, two electrons can never be in the same quantum state nor have the same set of quantum numbers, and therefore the electrons around an atom are treated as part of an entangled system. In the periodic system of the elements, higher elements add electrons in their orbitals in such a way that the periodicities of these become clear. Because of the nature of these orbitals and the electrons that fill them, the different types of atoms—nitrogen, oxygen, and so on—have different properties, or what we could call discrete "personalities," that tend to attract or repel other atoms with different "personalities." Based on combinations of atoms with these

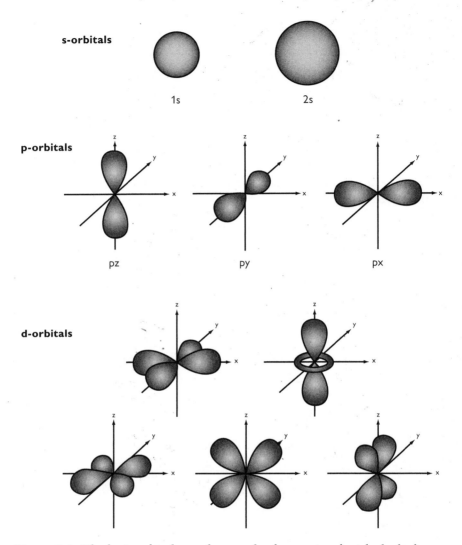

Figure 5.4. The basic orbitals, or electron clouds, associated with the hydrogen atom. The chemical elements of the periodic system is generated by a build-up of electrons with discrete orbitals in increasingly higher quantum states. Here the different quantum states and geometric forms of these electron orbitals will be looked upon as properties of the Atomic Tree of Life, a three-dimensional coordinate system. Because the electron energies around the Atomic Tree of Life are quantized, the electrons are not attracted into the nucleus, which would lead the atoms of the universe to collapse upon themselves. In addition, because the orbitals are quantized, there are no intermediate forms between the different chemical elements.

"personalities"—defined by different quantum numbers—molecules are formed. This gives rise to the chemistry from which matter, including in all biological organisms, is formed.

So far, so good! With these breakthroughs, quantum theory gained some excellent mathematical tools for predicting the behavior of light and subatomic particles. Yet when it came to what quantum theory actually meant, the notions began to diverge considerably and a number of different *interpretations* of quantum theory were proposed. According to the dominating Copenhagen interpretation—from Bohr and Heisenberg, who studied there—there was no such thing as "deep reality." In this interpretation, particles had no definite states until they were measured and many other absurdities that the aficionado of quantum theory is well familiar with. There are alternatives to this interpretation, however, such as David Bohm's "neorealism," which is less counterintuitive than the others. Instead of going into these discussions let us here just point out that the perceived "fuzziness" or absurdness of quantum theory is always associated with the fact that it has regarded the particle as the object of study and not the Platonic geometry of the atom. Thus while it may never be possible to know the exact location of an electron, or indeed if such a particle even exists, the *geometry* of its orbital is nonetheless very strictly defined and allows for an essentially determinist theory.

It deserves mention that, amazingly, modern physics provides no explanation as to why matter exists in an atomic form in the first place, something that may be food for thought for someone who thinks that science in its current form has all the answers. Even if the Schrödinger equation can provide exact descriptions of the shapes and sizes of atomic orbitals, it is very noteworthy that its underlying rationale is not specified nor is the very nature of the waves that it describes, which have no tangible physical meaning. Hence, the Schrödinger equation, the heart of quantum theory, has not been derived based on any stated physical principles. The reason is that modern physics has no theory as to why atomic matter exists in the first place, nor why it has the size that it does, and so there is nothing to deduce the Schrödinger equation from. It then becomes hard to understand how anyone could argue that the matter is primary to the geometry that constrains it. The Schrödinger equation is just a mathematical tool that has been shown to be useful for predicting waves, or orbitals, at different energy levels and quantum states, but it has no explanatory power.

In the Tree of Life interpretation of quantum physics, the invisible three-dimensional coordinate system (the Tree) around the atomic nucleus provides the Platonic framework for the orbitals of all atoms. We will here really turn quantum theory upside down and argue that the way subatomic particles behave is not a function of any inherent properties of these particles, but of the Platonic geometry that defines the extension of their orbitals. Naturally, a person with a materialist bias will object to such a view, but if the Schrödinger equation is not derived from any explicit physical reasoning, what scientific basis would there be to dismiss this view? *In the Tree of Life interpretation of quantum theory, the three-dimensional coordinate system that the orbitals of any atom are built around is looked upon as having microcosmic resonance with the Cosmic Tree of Life.* This would explain why atomic matter based on such orbitals exists to begin with. If the wavelengths of the orbitals throughout the universe are defined by an exact resonance with wave movements generated by the Cosmic Tree of Life, we would be able to understand not only why and how atomic matter is quantized, but also why these atoms, seemingly throughout the universe, have a uniform size and emit light with certain wavelengths. This provides a valuable analogy for how biological organisms are also primarily, as we have seen, defined by a Platonic geometry. In this new interpretation of quantum theory, reality is not seen as fuzzy, probabilistic, or nondeterministic, since such terms have no meaning when applied to the geometric shapes of the atom orbitals. Nor could it be said that "we create our own reality," as you sometimes hear said by popularizers of quantum theory. In the Tree of Life interpretation of quantum mechanics, the objective reality we live in, as well as our self-consciousness and perception of reality, is instead *created for us* by the Cosmic Tree of Life.

This interpretation would dispel some of the "fuzziness" and absurdities that are sometimes associated with quantum theory. It would also allow us to understand how the quantization of atomic matter, necessary for the emergence of life, could be generated by waves in resonance with the Cosmic Tree of Life. We may then ask if there is any kind of direct evidence that atomic orbitals are based primarily on a Platonic geometry generated by resonance with the Cosmic Tree of Life? The answer is yes. As I mentioned above, Bohr's atomic model is now considered outdated. Yet a significant parameter that has been retained from this model is the so-called Bohr radius. In the orbital model of electron clouds, this radius is defined as the distance from the nucleus at which there is the highest likelihood of finding the electron (in the lowest energy state in the hydrogen atom—see the s-orbital in figure 5.4). The radii of all other atoms may also be said to be related to the Bohr radius, for this the most basic quantum state of the hydrogen atom. The interesting thing is that this radius is inversely proportional to the strength of the electromagnetic interaction.* (This is easy to visualize. If this interaction was weaker, then the attraction between the electrons on the one hand and the protons in the nucleus on the other would be weaker, and the electrons' radii would be longer.) But according to the finding about electromagnetic interaction reported in the previous section, the Bohr radius is based on the golden ratio. We can then see that the size of the hydrogen atom (to which the sizes of all other atoms are related) and the very phenomenon of quantization is directly constrained by the phi constant. This constant is the chief spatial characteristic of the Cosmic Tree of Life. In the Tree of Life theory there is thus at least a tentative explanation as to why atomic matter exists and how the size of the Halos, which the Bohr radius describes, are mathematically defined by

---

*The Bohr radius can be calculated as $r_{Bh} = \dfrac{h}{2\pi m_e \alpha_{EM}} = \dfrac{h360}{2\pi m_e \varphi^2}$, where h is Planck's constant, m_e is the mass of the electron, α_{EM} the fine structure constant, and φ the golden ratio.

the Cosmic Tree of Life. In my view, this amounts to an enormous step forward for the new science about the living universe.

The role of the golden ratio in the geometry of atom orbitals however goes much further than this. It has been shown that the point of electric neutrality between the proton and the electron in the hydrogen atom is exactly defined by the golden ratio and that the radii of other atoms, as well as of molecules and ions, are also defined by this ratio.[19] The geometric properties of all atomic matter, including its molecular bonds, are thus constrained by the golden ratio, which indicates that they indeed are defined by resonance with the Cosmic Tree of Life. The reason that orbitals according to the Schrödinger equation would only be allowed to have certain quantized forms would then be that only some of their underlying wave forms would be in resonance with the Cosmic Tree of Life. With a few mathematical formulas it could also be shown that all light has an origin in this. Hence, not only are we human beings emitting light in a way that is defined by the Cosmic Tree of Life, but the whole universe may be looked upon as a great light show set up by the Cosmic Tree. With this shift in interpretation, quantum theory is transformed from a materialist to an idealist theory, quite consistent with the idea that the universe has a Platonic origin in a Big Thought that lies beyond material reality.

Based on the Tree of Life interpretation of quantum theory, where every atom is regarded as a three-dimensional Atomic Tree of Life, we may also understand in principle how the evolution of metallicity in the galaxy may have been synchronized with the evolution of cellular life. Such arguments of how the emergence of the physico-chemical conditions for life is synchronized with biological evolution are however now becoming repetitive, and so I will not discuss them in more detail. It is interesting to note however that developing an awareness of the Cosmic Tree of Life seems to dispel fuzziness. We have seen how a deterministic model consistent with the Tree of Life theory has come out of the study of biological evolution and dispelled its purported randomness. Now it seems that this same theory may also dispel some of the fuzziness and counterintuitive interpretations that have dominated quantum theory. The reappearance in our consciousness of the Cosmic Tree of Life tends to sweep away theories based on randomness, because it provides the universe with a unifying center in which evolution at all levels is anchored. The Cosmic Tree of Life in fact changes the entire context in which physical laws are formulated. We have found that the constants in some of the fundamental equations of modern physics are constrained by the Cosmic Tree of Life, and the reason is that the Cosmic Tree provides a completely new context in which these equations exist. In twentieth-century physics the mathematical equations used to describe the universe appeared to be valid regardless of any context and for no apparent reason. Like the Schrödinger equation, they just seemed to be useful and right by some strange accident, and scientists were not encouraged to explore their origin and purpose as long as they "worked." In the new paradigm presented here, mathematical equations describing the forces of the universe are instead right because they describe geometries and forces that are part of the context for life provided by the Cosmic Tree of Life.

6

The Wider Context of Biological Evolution

THE QUANTUM STATES OF THE COSMIC TREE
OF LIFE AND THE EVOLUTION OF THE UNIVERSE

*... flow is, in some sense, prior to that of the "things" that can
be seen to form and dissolve in this flow.*

DAVID BOHM

We saw in chapter 4 how biological evolution follows the quantized
Mayan calendar, and in chapter 3 how what we refer to as the physical
universe on different levels is fine-tuned to the evolution of life and in
fact is designed to generate life. The synchronized appearance of life-
creating phenomena on different levels of the Tree of Life also shows
that these levels are directly connected with one another, and this chapter will discuss how such synchronizations of such evolutionary processes are possible. To provide an understanding of this we will need to
present a new, broader context of the cosmos in which the evolution of
the biological species is promoted. This will require not only that we

study the relationship of biological evolution to galactic and geophysical evolution, but also that we grasp a couple of fairly difficult concepts.

The first concept is that the Cosmic Tree of Life, similar to the Atomic Tree of Life, may exist in different quantum states and brings about the evolution of the universe as it shifts between these. The second is that the lower levels in the hierarchy of life are *entangled* with the Cosmic Tree of Life, and brought to evolve through teleportation of quantum states. To propose that the entire cosmos exists in a quantized state, and that evolution is brought about by shifts between these states, could however very easily be misunderstood. Previously the consensus has been that quantum phenomena appear only at the atomic level or lower, and so I must emphasize that I am not implying that the reality on the higher levels can simply be reduced to quantum theory on the atomic level. In this theory it is the other way around: the quantum phenomena at higher levels are seen ultimately as defining those at lower levels.

Before we enter this discussion it may be relevant to ask why the universe evolves in the first place. We have here talked much about evolution, and biological evolution in particular, almost as if it could be taken for granted that our universe is an evolving universe. This, in fact, is how it quite commonly is looked upon in the modern worldview. Yet why is this so? The answer to this question is by no means trivial, and before we study the actual mechanisms of biological evolution in chapter 8, I feel it is appropriate to take a closer look at this. Could it be that the Cosmic Tree of Life may exist in different geometric energy configurations, in principle similar to the orbitals of the Atomic Tree of Life (figure 5.4), and evolve as it shifts between these configurations? If we look at the old Hunab-Ku symbol in figure 1.2, it seems that according to ancient myth this could very well be the case.

In support of this idea, we may take note of several fundamental similarities between atomic quantum theory and the Tree of Life theory: (1) The electron clouds (orbitals) are organized in atoms in a three-dimensional coordinate system (see figure 5.4), namely, the Atomic Tree of Life, much as we postulated such a system of axes for the Cosmic Tree of Life that gave rise to the big bang. The shapes and energies of these orbitals are quantized so that they can have only certain predefined values. (2) The wave movement defined by the Schrödinger equation lacks a direct physical meaning, which is true also for the wave movement generated by the Cosmic Tree of Life. The two therefore share what may be regarded as a philosophical problem (or solution, depending on how you look at it) that is of the same Platonic nature. (3) There is a periodic system of the elements corresponding to different energy levels much like there is a periodic system of biological evolution. Electrons may make quantum jumps between different orbitals in the three-dimensional system of the atom, similar to the way the shifts between Days and Nights may generate quantum jumps of new species. (4) When electrons in atoms shift between the quantum states of different orbitals, light is emitted, and when the Cosmic Tree of Life shifts between quantum states, Days (representing another kind of light) may begin.

If these analogies between quantum theory on the atomic and the cosmic levels are relevant, it would indicate that the quantum numbers of the Mayan calendar would, as previously suggested, correspond to different energy states of the Cosmic Tree of Life. Figure 6.1 shows hypothetical examples of different quantum states, with complementary, or opposing, yin/yang polarities. These are just examples, and when the 45-degree angles are included, similar to what we see in diagrams of atoms (figure 5.4), many other configurations are possible. What I propose is that the universe evolves because of such quantum shifts of the Cosmic Tree of Life. I also propose that the Days and the Nights of the Mayan calendar (and of *Genesis*) correspond to the quantum states of the Cosmic Tree of Life that generate wave movements, waves that indeed would also explain why the universe evolves in the first place.

I must state that there currently is no direct evidence that the Cosmic Tree of Life evolves through different quantized states (although if the preliminarily findings on the handedness of galaxies can be verified, this would provide evidence). However, in the periodic

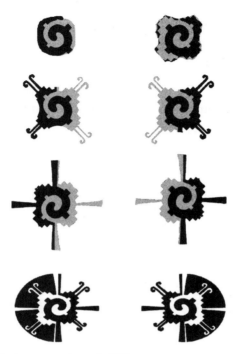

Figure 6.1. Symbols suggestive of different quantum states of the Cosmic Tree of Life, Hunab-Ku. The Hunab-Ku symbol for the Cosmic Tree of Life is here shown in different possible quantum states. The different states are discrete and do not shift from one to another in a slow continuous way, but as immediate quantum jump. Many different quantum states of the Cosmic Tree of Life are certain to exist. This model in principle is similar to the quantum states of the Atomic Tree of Life and its different orbitals, as seen in figure 5.4. The various axes of the Cosmic Tree of Life are vibrating with the periods of hablatun, alautun, kinchiltun, and so on of the Mayan calendar to create the various evolutionary wave movements of the universe. This is the reason that the Mayan calendar, unlike astronomically based calendars, describes a time that is quantized.

system of biological evolution in chapter 4, we saw that the time periods of the Mayan calendar are associated with quantum shifts apparently emanating from the Cosmic Tree of Life, and we have also seen that such quantum shifts are associated with the evolution of the body plan of species in three dimensions. This would mean that each species would be an expression of a discrete wave form associated with a certain three-dimensional quantum state of the Cosmic Tree of Life. If it seems strange to have an underlying Platonic geometry for all biological species, one may note that this also seems to be true for the geometries of the atomic orbitals. These geometries do not simply follow from the energies of the electrons.

The advantage of this model is that it would provide not only a common context for the Trees of Life at several different levels, but also help us understand why processes at these levels seem to be "collaborating" in bringing biological evolution about. In quantum theory this collaboration goes by the name of entanglement, which originally was seen as a controversial and inexplicable action at a distance. But according to Erwin Schrödinger, this is the most important aspect of this theory. Another advantage of a quantum model of biological evolution is that it may help explain the many quantum jumps that exist in the fossil record and why direct lineages of species are rarely, if ever, demonstrated.

The quantum numbers that would apply to biological evolution would then be exactly those that are associated with the number of Underworlds and Heavens that have passed in the Mayan calendar, which correspond to the dates associated with quantum leaps in evolution. These quantum numbers are not the same as those of the periodic system of the elements, but the system would in principle be similar to it. In the Tree of Life interpretation of quantum theory, the quantization of physical reality would not be limited to its subatomic or atomic levels, but apply to higher levels of the cosmos as well. If the atom as well as the cosmos at large are both expressions of the Cosmic Tree of Life, why would quantum states be evident on only one of these levels?

What then is the evidence that the higher levels of the Cosmic Tree of Life are quantized? Part of that evidence is in the form of the quantized biological evolution we discovered in chapter 4. Part of it comes from human history, and in earlier books I have shown how the Planetary Tree of Life has geographic effects that are clearly quantized in the higher Underworlds. Paradigm shifts in human thinking, which we will see a single example of shortly, in fact often correspond to quantum jumps correlated with the Mayan calendar.

At present it does not seem possible to propose exactly what the quantum states on the level of the Cosmic Tree of Life might look like, but only to propose a model for this in principle. In my previous books I presented a Cosmic Round of Light, more in the language of the ancients than of quantum theory, to explain shifts in evolution. This will not be reproduced here except in figure 6.3 (see page 164), where we can see how different quantum states manifest in different yin/yang polarities. In the columns to the right in figure 6.3, we can see how these affect the Earth and its biological species (symbolized by a human head) in the different Underworlds.

Thus we can understand the origin of the wave movements of the Underworlds if we regard an Underworld as the outcome of a periodic induction of a separation of opposites, or dualities, corresponding to different quantum states. The introduction of a new quantum state results in a new duality, and in the emergence of a creative tension between yin and yang (see an example on the level of the Earth in figure 6.2a). This in turn manifests as an opening for novelty until unification, and a new balance, is attained between yin and yang when this temporarily returns to a non-polarized quantum state (figure 6.2b). The source of these separations of opposites, these yin/yang polarities, is the Cosmic Tree of Life, which shifts between quantum states in relation to its axes in three dimensions. In figure 6.1, a more complex model shows these axes are complemented with eight-partitioning ones, although it may be noted that in reality the Cosmic Tree of Life is three-dimensional. The axes of the Cosmic Tree of Life may be seen to serve as wave generators by periodically, with the given rhythm of an Underworld, introducing a boundary between yin and yang (which also means shifting quantum states), and it is through the introduction of this boundary that the polarity between the two is created. When a yin/yang polarity is introduced along one of the axes of the Cosmic Tree of Life, a Day begins, and when the polarity comes to an end, a Night begins (see an example

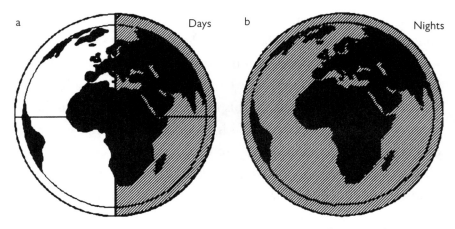

Figure 6.2. The Days and Nights are dominated by different Halos. During the Days of any Underworld, the Halos create a yin/yang-polarized field, whereas during the Nights, the Halos are dark. In any Underworld, a wave movement between polarity and unity is created by the alternations between these two types of Halos. These Days and Nights reflect alternating quantum states of the Cosmic Tree of Life. Here the generation of the wave movement of Days and Nights is exemplified at the level of the Earth (although the same principle applies synchronistically to all other levels as well; see figure 6.5). The nature of the yin/yang polarity during the Days however varies from Underworld to Underworld (figure 6.3) according to the Cosmic Round of Light. The example shown here is from the Mammalian Underworld.

on the level of the Earth in figure 6.2). In this way we may understand how a wave movement of alternating Days and Nights is created by the periodic introduction by the Cosmic Tree of Life of a yin/yang polarity about its axis. Thus when this boundary is introduced on the level of the Halo, it generates a time period, a Day, when the yin/yang polarity creates novelty, while when the Cosmic Tree of Life deactivates the boundary, a period of lesser creativity, a Night, begins, which through its unitary character serves to integrate the novelties of the previous Day and balance them in preparation for the next Day.

To explain that evolution has a different character in each of the different Underworlds, it is also necessary to say that the yin/yang polarities, or quantum states, are different in the various Underworlds.

The Cosmic Round of Light provides a description of the quantum states underlying the periodic system of evolution. In this the boundaries of the yin/yang polarities shift 90 degrees clockwise, when looked upon from above (see figure 6.3), each time the evolution of a new Underworld is activated. With every new Underworld a new wave movement of seven Days and six Nights begins, which has a yin/yang polarity that is shifted 90 degrees from to the previous. As this is completed a new wave movement with a new yin/yang-polarity is activated, and so forth, until after several such steps the top of the pyramid is reached.

Incidentally, in ancient myth you often find a serpent, or dragon, associated with the Tree of Life (figure 6.4). This is because the serpent is a symbol of the wave movement that emanates from the Tree of Life as this shifts between quantum states. In fact, the oldest known piece of art on Earth is a seventy-thousand-year-old serpent carved in rock in Botswana,[1] and according to the San people of this region, who also included the Tree of Life in their mythology, humankind descended from a python. If the python is interpreted as a wave movement, this makes perfect sense, since in the quantum model the human being is a wave form, a Platonic wave form if you like.

Because in all of the Underworlds the quantum states of the Cosmic Tree of Life create a polarization between yin and yang, the universe is always out of balance and was so from its very beginning. *It is this imbalance on the level of the Halo, and the shifting quantum states of the Cosmic Tree of Life, that drives the material universe to evolve.* The shifting imbalances in different Underworlds make this evolve in a directed and essentially orderly way despite occasional appearances to the contrary, and without these imbalances there would be no evolution. A result of this inherent imbalance, incidentally, is that the universe has never been perfectly symmetrical, but always includes a measure of asymmetry within the symmetry. The bodies of higher animals, including our own, which have been created in the image of this polarized cosmos, also to a degree reflect these cosmic asymmetries, such as in the functions of brain hemispheres. In my previous

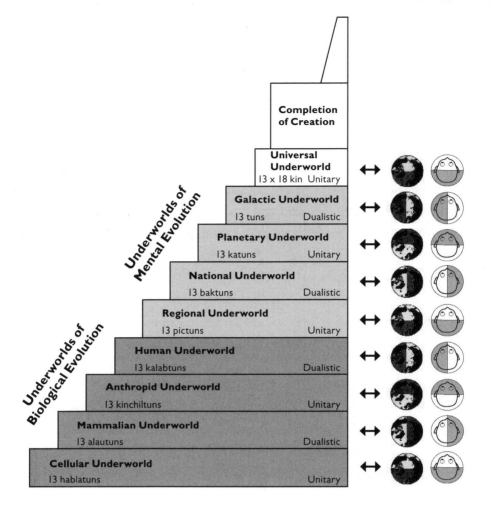

Completion
of Creation

**Universal
Underworld**
13 x 18 kin Unitary

Galactic Underworld
13 tuns   Dualistic

Planetary Underworld
13 katuns   Unitary

National Underworld
13 baktuns   Dualistic

Regional Underworld
13 pictuns   Unitary

Human Underworld
13 kalabtuns   Dualistic

Anthropid Underworld
13 kinchiltuns   Unitary

Mammalian Underworld
13 alautuns   Dualistic

Cellular Underworld
13 hablatuns   Unitary

Underworlds of
Mental Evolution

Underworlds of
Biological Evolution

Figure 6.3. The Cosmic Round of Light. The yin/yang polarities ruling the Days differ in character in different Underworlds. As the Cosmic Pyramid is climbed, these yin/yang polarities on the level of the Halos shift 90 degrees with each higher Underworld. This pattern applies synchronistically to all the different levels of Halos, although to the right here only the Earth and the human being are used as examples of the effects of these shifts. These are shown from the north pole and the top of the head, respectively. If you use the direction of vision of the human beings as the point of reference, it is possible to see how the shifting yin/yang polarities alter this perception and creativity. Some Underworlds will be dominated by a left-right polarity and others by a front-back polarity. It should be pointed out that in the lower Underworlds, the Earth may not have looked as it does now and lower organisms should have replaced the human being in the model (because humans did not exist until the fourth Underworld). The purpose here, however, was simply to show the principle of the Cosmic Round of Light.

books I have extensively discussed how these changing yin/yang polarities may explain some very significant aspects of the mental evolution of humankind in four of the Underworlds (see figure 6.3). The yin/yang polarities determine if the emphasis in an Underworld is on the expression of the left or of the right half, or front or back, of the brain. Through our resonance with the Cosmic Tree of Life, human beings perceive the world through the mental fields created by its different quantum states, and the shifts between these are what create paradigm shifts in science and elsewhere.

Because this topic has been discussed earlier, I will only repeat a few observations that are especially relevant here. First, the design of the nine Underworlds seems to be that the climb of them is completed with a field of Light, dominating our front side in the Universal Underworld.

Figure 6.4. Tree of Life in Codex Selden. The birth of a man (called Two Grass) from the Tree of Life, with serpents corkscrewing around it in a Mixtec codex. Trees of Life in many traditions had serpents associated with them because serpents are symbols of the wave movements that emanate from the Cosmic Tree of Life. Note also the curious eye in the middle of the tree, and compare this to the Mayan Tree of Life in figure 1.3. (Page 1, Band 2, Codex Selden.)

If the plan works out, a perception of the unity of all things would be generated, which could be described as a return to the Garden of Eden–like Regional Underworld, albeit at a much higher level. Second, the reason we today need to go back to ancient science, such as the Mayan calendar, to gain a deeper understanding of the evolution of the universe is that ancient science was generated through a different yin/yang polarity of perception (see the fields of human perception of the head to the right in the National Underworld in figure 6.3). The ancients saw things that for the most part we have not been able to see in the Planetary Underworld, since each Underworld produces its own thought structures and its own science based on the particular yin/yang polarity in the Cosmic Round of Light that has generated them.

While the science of the National Underworld (with spiritual light on the left brain half) recognized an objective reality in absolute time and space, this came to be denied in the observer-centered physics of the Planetary Underworld (with no spiritual light on either brain half). As we noticed in chapter 3, around the time of activation of the higher rhythm of the Planetary Underworld in 1755, the earlier view of a static universe was replaced by one of evolution. Maybe it is only now, with our entrance into the Galactic Underworld, that the nature of this evolution can be understood. In the Galactic Underworld (with spiritual light on the right brain half), which we entered in 1999, a holistic understanding of the evolution of the cosmos as intelligently designed with an existence independent of the observer is again favored. Because of the shift in perception that this entails, we are starting to recognize an objective reality created by the space-time organization emanating from the Cosmic Tree of Life that is not limited to the atomic and subatomic levels, such as in traditional quantum physics.

The proposal that the Cosmic Tree of Life changes quantum states is, needless to say, very radical. Nonetheless, it is not completely unprecedented, as the big bang is sometimes in established physics referred to as a quantum event. What I have done is simply to expand from this original quantum event to explain the many quantum jumps that followed in biology and history. The case for quantized evolution becomes even stronger when we look at entangled galactic and geophysical levels.

ENTANGLEMENT

This section discusses the use of quantum theory to understand how the different levels of organization of life are synchronized to support biological evolution. In chapter 5 we saw how the evolution of metallicity (which really is an expression of the build-up of higher quantum states in Atomic Trees of Life) was synchronized with the emergence of cells (which is an expression of higher quantum states of the Cosmic Tree of Life). In quantum theory synchronized events may be explained by a phenomenon called entanglement, which is defined by Wikipedia as follows: "Quantum entanglement is a quantum mechanical phenomenon in which the quantum states of two or more objects are linked together so that one object can no longer be adequately described without full mention of its counterpart—even though the individual objects may be spatially separated. This interconnection leads to correlations between observable physical properties of remote systems." In other words, certain properties of particles or objects become intrinsically connected if they are created together or interact with each other. It is because of entanglement that a particle seems to "know" what state another particle is in if the two are part of the same system. For instance, because they are entangled, two different electrons around an Atomic Tree of Life never have the same spin states or quantum numbers. Somehow each electron seems to "know" what states the other electrons in the atom are in. Because of the entanglement of electrons, the Schrödinger equation treats the wave form of an atomic system as a whole and not as a sum of the waves of the individual electrons.

For a long time scientists debated whether particles would remain entangled at very large distances from one another when no classical means of communication between them seemed possible. In the 1960s it was however proven by CERN physicist John Bell that entanglement was nonlocal. In fact, we are living in a nonlocal universe where objects could be entangled over vast distances without any energy exchange taking place between them. This implies that the spin of a particle may be connected to the spin of another in such a way that if the spin of one of them changes then *the spin of the other will change synchronistically regardless of the distance between them.* This was later proven by French physicist Alain Aspect in an experiment that has been heralded as one of the most important in twentieth-century physics.

What is different in the present theory is the radical proposal that quantum phenomena may exist on several levels of the universe, not only the atomic or subatomic. In developing this theory, it is often the connections between the different levels of the Cosmic Tree of Life that need to be explained. What we will see in the following two sections is how Halos and their spins—also at different levels of the universe such as the galactic, planetary, and biological—may be entangled and undergo shifts synchronistically at critical points in the Mayan calendar. This is because the Trees of Life on these different levels indeed seem to be entangled. At least on the level of atoms it is well-known that quantum states can be teleported,[2] and this may be at the root of the coordination of the evolution of the universe. Synchronistic changes at different levels of the universe, which play an important role in the Tree of Life theory, may thus find their explanation in entanglement between these different levels. This entanglement of wave forms, or Halos, may in fact well be what Bohm referred to as the implicate order of the universe.

Quantum theory postulates that *certain* properties, such as spin, of any two particles or objects are entangled if they were created together or interact with one another. It has been argued that the entire universe is entangled, since all of its matter was formed together in the big bang. Such an entanglement of everything with everything else however seems like an oversimplified and incorrect view. If I for instance raise my arm, others in the universe will not be raising their arms synchronistically, and the spins of the electrons in the arms would most likely not be shifted synchronistically either. Early on in chapter 1, however, I postulated that the big bang was primarily an organizing event, which established a seniority rule between different levels, and I believe that ever since certain properties in certain systems have been entangled. For instance, in my previous books, I present several examples of how the mind of the human being is entangled with the yin/yang polarities (quantum states) of the Earth (a single example was given in the previous section about the history of science). In the following sections we will see further examples of entanglement of Trees of Life at different levels of the universe and the expressions of these in their respective physical systems.

As mentioned, a way of looking at entanglement is as a teleportation of quantum states. This is a phenomenon that currently plays a big role in attempts to develop quantum computers. What it means is that a state at a point A can be made to appear at a distant point B without actually traveling the distance between the two points. To understand how the universe is created and evolves, we may look upon the Cosmic Tree of Life as a giant quantum computer, which teleports quantum states, synchronized on different levels, across the universe. That indeed the different levels are synchronized at such cosmic quantum shifts is something we will see examples of in the following sections.

Before we continue I feel it is pertinent to respond to the possible objection of some critics of the present theory, who might consider Halos, or wave movements, as entities of a "supernatural" character and hence argue that they have no place in science. My own view is however that it only adds confusion to an issue to call a phenomenon "supernatural." Either a phenomenon exists—and should therefore be called natural— or it does not exist, in which case there is really no point in discussing it at all. We may also note that the philosophical problem of the Tree of Life theory is identical to that of the established quantum theory of particle physics. This problem is that the wave forms in both theories lack an apparent physical meaning. It however often happens in science that the existence of a field needs to be postulated because its effects are observed, and in this regard the Tree of Life theory is no different. In line with this, the reason that the existence of the Halos, or quantum states, needs to be postulated is that we see their effects, and that the emergence and evolution of the universe cannot be coherently explained without the existence of such wave forms.

A special characteristic of the wave forms created by the Cosmic Tree of Life at higher levels of organization is that they undergo periodic changes with the different rhythms described by the Mayan calendar. The fact that they bring about the evolution of the universe according to a given time plan, and that this was mythologized in ancient times, does not make them different from any other fields of physics, except that they may play a more important role for how we perceive the purpose of the universe and our own roles within it. The Halos, in other words, are what ultimately drive the evolution of the universe forward and gives it its direction. Because of these existential consequences, the Halos may create more controversy, and possibly discomfort for those wedded to the notion of a pointless universe, but this is hardly a scientific argument against their existence.

We should also note that if the existence of wave forms was demonstrable, that is to say, if they could somehow be influenced by matter or energy, then the existence of life, and the whole nested organization of the universe into distinct levels, would be immensely more precarious than it actually is. The coherence of the various organizations of life would constantly be under threat of disintegrating if the Halos were subject to physical influences, and life would not have been particularly likely to emerge anywhere. As it now stands the Halos do not seem to be affected by physical conditions, since they are its primary space-time organizers and are senior to energy and matter, like an evolving Platonic matrix of the universe. They provide the primary structures that promote the emergence of life in the universe. A consequence of this is that Halos as such cannot be manipulated or experimented with by human beings and that we therefore are not able to create life, but only sometimes to modify it when it already exists.

I should also add that the present model of entangled Trees of Life at different levels is consistent with the existence of several phenomena that have qualified as paranormal, such as telepathy. If Halos on different levels are entangled, this means that the minds of different people may be entangled with a Halo on a higher level and may therefore experience synchronistic phenomena such as telepathy. Usually such effects seem very difficult to demonstrate, presumably because the intention to have a paranormal experience interferes with the possibility of actually having one. Rupert Sheldrake however has indisputably demonstrated telepathic phenomena between pets and their owners,[3] presumably because in these cases the pets are completely free from any intention of experiencing them. The existence of a Platonic matrix of wave forms behind the solid reality of the universe also opens up the possibility of understanding the "world beyond" that is contacted by psychics.

THE SEXUAL POLARITY OF THE GALAXY

We have now covered some basic ground for the development of a new theory for the evolution of the universe. The cosmological principle and Darwinism have proven to be inadequate, if not outright wrong, and a way has been pointed out for the development of a new framework of quantum theory that is consistent with the new findings. In fact the falsification of Darwinism directly mandates that a new theory of biological evolution be developed, and the rest of this book will be devoted to this and some of its existential consequences. We have seen that the generation of new species of organisms is not the result of a purported "struggle for survival," but is an integral part of a larger fine-tuned cosmic scheme that creates wave movements and "quantum jumps" to new species, in principle not unlike what quantum mechanics describes on an atomic level. In this scheme, the evolution of the universe is driven by shifts between quantum states of the Cosmic Tree of Life, and as an aspect of this, biological evolution is developed by the four lowest Underworlds. It has been suggested here that the emergences of biological species are effected according to a higher purposeful time plan in accordance with the Mayan calendar. The execution of this time plan requires the entanglement of several different levels of the universe and the generation of synchronistically shifting wave forms on these levels.

We have already found that wave movements originating from the Cosmic Tree of Life create periodic transformations of biological organisms. Yet we have seen little of how these wave movements are mediated through the hierarchy of Halos and how they are received and expressed in the creation of new species. The point of departure for such an exploration of the mechanisms for the transmission and expression of morphogenetic fields must obviously be the Cosmic Tree of Life. Over a few billion years this has generated not only single cells, but also multicellular organisms with vision and lateralized brains having the power to process sensory information, and eventually an erect species, the human being, as the most advanced reflection of the Cosmic Tree of Life. We then need to study how the wave forms are related to the hierarchy of space-time organizers that the Trees of Life on different levels (figure 1.5) constitute and how these contribute to the creation of biological evolution. Indeed, as we have already seen, the evolution of biological organisms seems to require such an entangled hierarchy of space-time organizers operating at different levels of the cosmos, since it is difficult to see what else would allow for their synchronization with the evolution of the physico-chemical conditions for life. Most of the remainder of this chapter will be devoted to studying the effects of the entanglement of Trees of Life at different levels.

We will here focus on how biological evolution is related to and synchronized with the evolution taking place at higher galactic and planetary organizations of life (figure 6.5). As we saw when we studied "dark matter," the structure of a galaxy seems to be organized by a Halo with a spinning axis, a Tree of Life, at its center. If this is so, we may wonder if this Halo and its yin/yang polarity plays any further role in providing a framework for the evolution of life and if it shows any evidence of following the Day-

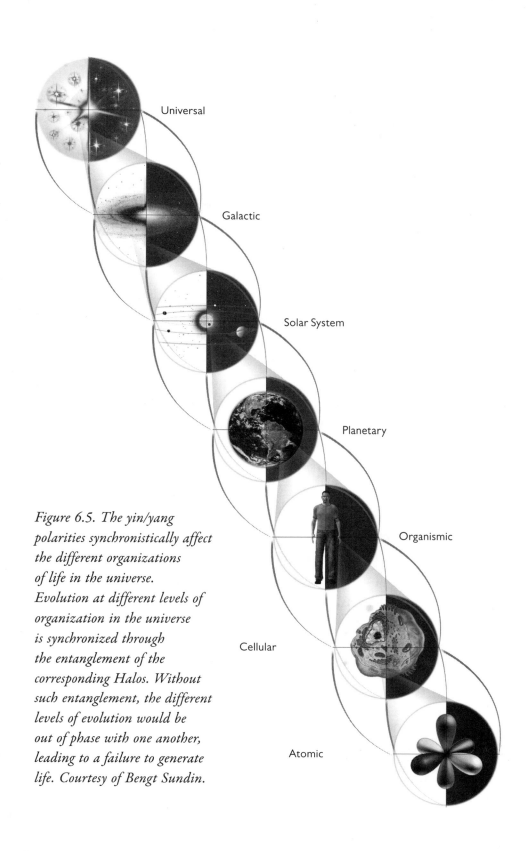

Universal

Galactic

Solar System

Planetary

Organismic

Cellular

Atomic

Figure 6.5. The yin/yang polarities synchronistically affect the different organizations of life in the universe. Evolution at different levels of organization in the universe is synchronized through the entanglement of the corresponding Halos. Without such entanglement, the different levels of evolution would be out of phase with one another, leading to a failure to generate life. Courtesy of Bengt Sundin.

Night wave rhythms of the Mayan calendar. One interesting example of such a rhythm is that the solar system, as well as the vast majority of stars in the galactic disc, moves up and down through the equatorial plane of the galaxy in a sine wave whose period is approximately 62 million year long (see figure 6.6).[4] The estimate of this period is very close to a Mayan alautun of 63.1 million years, meaning that during a galactic year (the duration of a revolution of a star around the galactic center) of about 250 million years,[5] the solar system completes approximately four cycles about the galactic midplane, resulting in a quadripartitioning of this plane. At the current time our solar system is relatively close (68 light-years) to this midplane of the galaxy, but it is moving away from this midplane toward the northern hemisphere of the galaxy.* About 3 million years ago the solar system thus passed the equatorial plane of the galaxy and will not again be aligned with it for another 59 million years.[6]

The most likely explanation for this cyclical movement up and down through the midplane of the galaxy is that the induction of yin/yang polarities generate this during the alautun-long Days of the Mammalian Underworld. Hypothetically then the tension generated by the yin/yang polarities of the galactic Halo along its midplane would cause the star systems to move away from the galactic midplane, and the gravitation of the galactic disc would cause them to move toward it, so that their combined effects generate a cyclical movement about this plane. Since we have already seen that the movements of stars in a galaxy are primarily determined by its "dark matter" Halo, it is not very farfetched to assume that the tension generated by the periodic introduction of yin/yang polarities of this very same Halo would have the power to create such a sinusoidal movement.

---

*It is today often said that on December 21, 2012, when some believe the Mayan calendar comes to an end, the Earth will pass through the galactic midplane (see for instance Dan Eden, "The Real Doomsday? December 21, 2012," www.mondovista.com/endtime .html). This view is not correct, and for the past 3 million years the solar system has been moving away from the galactic midplane. It will continue to do so for some 28 million more years.

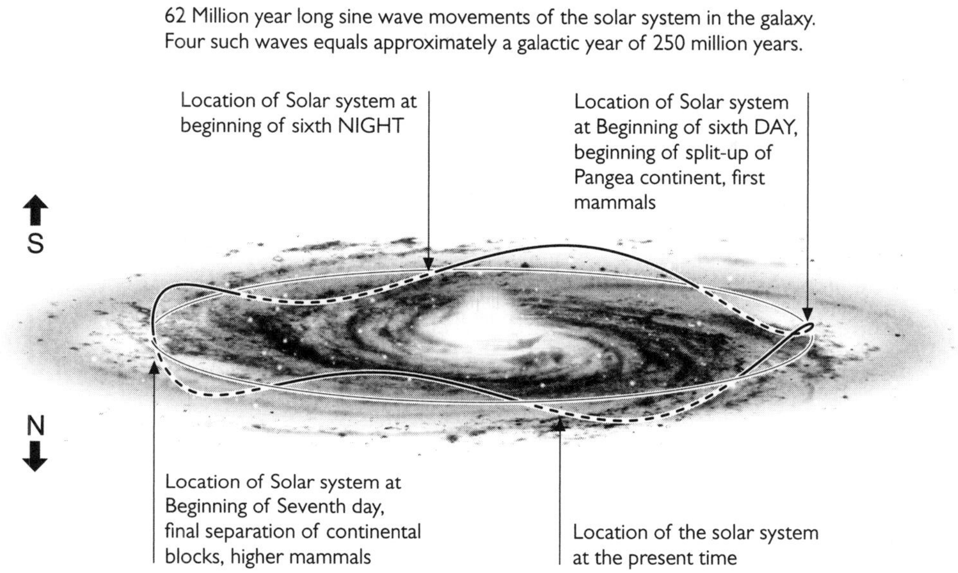

62 Million year long sine wave movements of the solar system in the galaxy.
Four such waves equals approximately a galactic year of 250 million years.

Location of Solar system at
beginning of sixth NIGHT

Location of Solar system
at Beginning of sixth DAY,
beginning of split-up of
Pangea continent, first
mammals

S

N

Location of Solar system at
Beginning of Seventh day,
final separation of continental
blocks, higher mammals

Location of the solar system
at the present time

Figure 6.6. Sine movement of the solar system in the galaxy. The solar system, as well as the vast majority of stars in our galaxy, displays a sinusoidal movement up and down through the equatorial plane of the galaxy in which each sine wave has a duration of about one alautun. This means that shifts between Days and Nights happen approximately four times in a galactic year of 250 million years and that significant events in terrestrial evolution will be correlated with this wave movement. At the current time we are moving away from the galactic equator, and the solar system will not be aligned with the galactic midplane for another 59 million years. Note: In the figure the amplitude of the wave movement of the solar system has been exaggerated.

In the presentation of the Cosmic Round of Light it was postulated that the introduction of yin/yang polarities on different levels, including the galactic, was the driving force behind evolution. In line with this we can see in figure 6.3 that in the Mammalian Underworld the polarity takes the form of a left-right polarity, which in figure 6.5 is shown at all Halo levels of the nested hierarchy. Maybe then there is a direct relationship between the previously discussed pulse-wise emergence of classes of species with a rhythm of 62 ± 3 million years and this sinusoid movement about the galactic midplane. Since the present theory postulates that the different levels of organization of life are entangled and so undergo change in synchrony at calendrical shift points, such a relationship would seem to be a logical consequence of the theory. The most obvious characteristic of the evolution of animals Day by Day in this Underworld is the stepwise development of the increasingly more advanced bilateral central nervous systems for processing visual information. An evolution in the direction of increasingly more advanced bilateral brains is an understandable effect of the introduction of yin/yang polarities in the Halos of the Mammalian Underworld.

The lateralization of human brains into an analytical left half and a holistic right half (in right-handed people) is well known. In most animals more basic biological processes are also affected. For instance, it is common, at least in birds and mammals, for expressions of sexuality and aggression to be preferentially mediated by the right brain half.[7] Given that the cosmos seems to be lateralized, such lateralizations on the part of animals do not seem surprising in a universe whose purpose is to generate organisms in its image. This would indicate that the sinusoidal movements of star systems and the evolution of classes of animals with lateralized brains have a common origin in the introduction of yin/yang polarities, or new quantum states at times when Days begin in the Mammalian Underworld. This is not to imply that one of these phenomena causes the other. Instead, to explain such synchronicities I propose that Halos at different levels of the cosmos are entangled such that yin/yang polarities are introduced synchronistically and have parallel effects on different levels, in this case on the galactic and the organismic, so that hemispheric polarizations are created on both of them.

For this model to be applicable we would have to assume that the galaxy is polarized into two hemispheres so that one corresponds to the left and the other to the right hemispheres of an animal brain. There is currently no evidence for (or against) any such polarity, but since both the cosmos at large and the brains of vertebrates are polarized, it would be more of an anomaly if the northern and southern hemispheres of the galaxy were not polarized. This would also mean that inhabited planets above and below the midplane of the galaxy might be under different influences of the yin/yang polarity. Even though all star systems in the galaxy move up and down between the two hemispheres, the nature of life on any one of them may be most strongly influenced by the hemisphere it is currently located in. The galaxy may thus be divided, polarized, into an upper and a lower domain, somewhat similarly to brain halves. Through this organization of the galaxy, multicellular organisms, including human beings (and extraterrestrial cousins), would synchronistically be affected by the same yin/yang polarities, and evolution would hence follow the same rhythm in all the star systems of the galaxy. The galaxy would serve as a womb in which planets with life are created in the encounter between the male and the female energies of the galactic hemispheres. The Desana people, an indigenous people in Colombia whose shamans hold significant knowledge of the human brain, including awareness of its lateralization, incidentally, believe that the galaxy is a "cosmic brain," separated into two hemispheres by the Milky Way.[8]

As mentioned above, the sexual responses in the brains of animals seem to be lateralized, and so we may wonder if the yin/yang polarities introduced on a galactic scale also have something to do with the generation of the two different genders. That there are two genders may seem obvious at first, but there is actually no theory that explains why higher animals have two and only two genders, and it is rare that the question is even asked. The Neo-Darwinist explanation for the existence of two biological genders is that their reproduction has the advantage of producing a greater genetic variability, but this no longer seems relevant, and in the Tree of Life theory there is no particular advantage for a species to generate genetic variability. If the need for such genetic variability were the real reason that there are two genders, it would seem it would have been even more advantageous to have three or even a hundred genders involved in reproduction. We may also ask why higher animals are divided into genders in the first place, since virgin birth is a biological possibility that is practiced in lower species, where it goes by the name of parthenogenesis. This would in principle be a much more reliable and effective means of reproduction, since there would be no need to find a partner of the opposite sex to reproduce. It seems that a better explanation for the existence of two and only two genders among higher animals is that they are generated by the fundamental yin/yang polarity of the universe and in particular on the galactic level. In support of this possibility we see in table 6.1 how the polarization into different genders has evolved step-by-step during the different Days of the Mammalian Underworld.

TABLE 6.1. THE EVOLUTION OF SEXUAL POLARITY IN BIOLOGICAL ORGANISMS IN THE MAMMALIAN UNDERWORLD

Heaven (Day or Night)	Beginning (Myr)	Class of Organisms	Mode of Reproduction
Day 1	820.3	First multicellulars	Splice off single cells?
	757.2		
Day 2	694.1	Ediacaran Hills fauna	?
	631.0		
Day 3	567.9	Trilobites	Lay eggs
	504.8		
Day 4	441.7	Fishes	Lay eggs and roe; no parental care
	378.6		
Day 5	315.5	Reptiles	Internal fertilization; lay eggs and give limited care to offspring
	252.4		
Day 6	189.3	Proto-mammals	Give birth to live offspring and nurture them in pouch
	126.2		
Day 7	63.1	Higher mammals	Give birth to live offspring and care for them a long period

It is important to note that the Cellular Underworld was not dominated by a lateralizing yin/yang polarity (instead, we find one between the front and the back; see figure 6.3) and that this Underworld thus did not generate two different genders. In the model we are presenting here a differentiation into two genders is thus in the Mammalian Underworld paralleled by the lateralization of the brain. Since we may infer that yin/yang polarities are introduced on the galactic level (from the sinusoidal movement with an alautun rhythm of the solar system about the galactic midplane), we also have some evidence that these yin/yang polarities are entangled with those on the organismic level. Hence, the increasing polarization into genders as new Days begin in table 6.1 is likely to be entangled with those on a galactic level.

The separation into two genders in the Mammalian Underworld would then be explained by the introduction of lateralizing yin/yang polarities on several entangled levels. *The hemispheric polarization into yin and yang would explain why there are two and never more than two genders.* The two genders would also be separated by a difference in resonance with different hemispheres of the galaxy, both longing for wholeness and union. Thus, *sexual reproduction in animals would have evolved in response to a striving to bridge the duality introduced by the Cosmic Tree of Life on the level of the galactic Halo.* Sexual activity in humans then also means a possibility of bridging the basic duality of the cosmos and its separation into yin and yang. From this viewpoint, *sexual desire in animals and human beings has a truly cosmic origin* and is only partially based on a desire for pleasure. Its origin is ultimately the drive to transcend a cosmic duality and it thus has heavenly dimensions. This would be the real reason that sexuality has been considered sacred in many spiritual and religious traditions. The attraction that it creates forms the basis of the formation of families and the long periods of care for offspring among human beings. It is questionable if any social life would have evolved among humans if there were not a sexual drive to bridge a cosmic duality. Ultimately then the sexual polarity goes back to a polarized quantum state at the level of the Cosmic Tree of Life with which the galactic level is entangled.

THE CONTINENTAL DRIFT AND THE LATERALIZED BRAIN IN A GALACTIC CONTEXT

We have now seen an example of how the galactic and organismic levels seem to be entangled and undergo shifts in synchrony. As an alternative perspective we may look upon the galaxy as a lens for Halographic information that is transmitted to the level of biological organisms. Another potentially very important lens is provided by the planetary system of Earth, which in countless ways appears to be designed for the emergence and evolution of life, most profoundly through its resonance with the tun orbit. Another very interesting way in which the Earth has been prepared for the emergence of life is through its plate tectonics, or continental drift as it was previously called, which refers to the movement of segments of the Earth's crust on the underlying mantle. Among the planets in our solar system, Earth alone manifests such phenomena.

As is argued in detail in *Rare Earth* by Peter Ward and Donald Brownlee at the University of Washington, it is to plate tectonics that we owe not only the existence of the continents that we inhabit, but also an atmosphere with carbon dioxide that gives the Earth a temperature well-adjusted for higher life.[9] There are too many important aspects of how plate tectonics makes our planet habitable, with oceans and continents existing in a reasonable balance, for them to be described here. Suffice it to say that the current structure of the Earth's crust goes back to this recycling of continents, and that in its first billion years the Earth was essentially a water world where human life would not have had much space to develop. We may then ask if this state of affairs is merely fortuitous or if the evolution of the continental structure of the Earth can be understood in terms of the Mayan calendar—the time plan of evolution. In other words, is it possible to link continental drift to quantum shifts in the Cosmic Tree of Life?

As a first indication of such a connection between plate tectonics and the Cosmic Tree of Life, there is weak evidence that sea level changes follow an alautun periodicity.[10] This is of interest because sea level changes obviously are a result of the effects of plate tectonics on the continental structure. Moreover, certain geologically defined eras, such as Carboniferous, Devonian, and Ordovicium-Silur, presumably arose from cyclical shifts in plate tectonics, each lasting approximately an alautun, which again indicates that plate tectonic movements follow this periodicity. Computer simulations of the early continental structure of the Earth indicate that its major blocks were separated during Days 3, 4, and 5 of the Mammalian Underworld.[11] In one study, a supercontinent is described to have existed at the beginning of the Mammalian Underworld, 750 million years ago,[12] but since this time the North American continent has had a "nomadic" character in which it has alternately been separate from and unified with the other continental blocks of the planet.[13] This would be a typical effect of the lateralizing yin/yang polarities of the Mammalian Underworld on the Earth level (figure 6.3). Despite the scant data there are thus indications that even in the distant past continental blocks may have been markedly separated during the Days of the Mammalian Underworld.

All knowledge about changes in the continental structure of the Earth happening more than 250 million years ago is very uncertain. At that point in time, however, it is known that the major continents of the planet, which during Day 5 (315–252 MYA) had been separate, merged to form the so-called Pangaea continent, a continent that remained in existence for the entire period of Mayan Night 5 (252–189 MYA, approximately the Triassic period). Around the beginning of the sixth Day (189 MYA), this supercontinent started to break up (see figure 6.7; the oldest sediments on the Atlantic sea floor have been estimated to be 180 million years old[14]). At the beginning of the seventh Day (65 MYA), the current main continental blocks, the Americas and Eurasia, were completely separated as Greenland started to drift away from Scandinavia and the North Atlantic was formed. As expected, the beginnings of both the sixth and seventh Days of the Mammalian Underworld, characterized by lateralizing polarities on the Halo level, led to a definitive separation of what had earlier been the Pangaea continent.

Continental drift

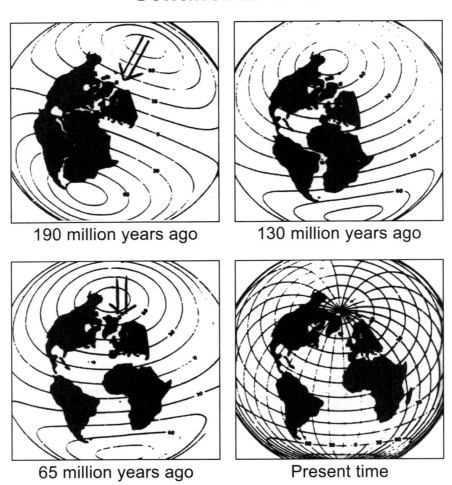

Figure 6.7. *The changing continental structure of Earth in the Mammalian Underworld. The Pangaea continent that had formed about 250 million years ago started to separate into two continental blocks about 190 million years ago at the beginning of the sixth Day. The complete separation between Eurasia and the Americas began approximately 65 million years ago around the beginning of the seventh Day. The introduction of yin/yang polarities at the beginning of Days thus gives rise to the separation of Earth's continental structure into two blocks. Used with permission from Bengt Loberg, Geologi, Norstedts, Stockholm.*

Going back to chapter 4, we may also note that in the Mammalian Underworld it is the Days that stepwise give rise to classes of species with increasingly more advanced lateralized brains. Given the above this suggests that there is a parallel between the drifting apart of the continents of the Earth and the emergence of biological species with increasingly more advanced lateralized brains. This becomes evident if we study the parallelism between the stepwise evolution of the mammalian brain and the continental drift, which we can do by comparing table 4.2 and figure 6.7. At the beginning of the sixth Day, when Pangaea was starting to drift apart, the first mammals appeared, and at the beginning of the seventh Day, when the continental blocks of the planet separated completely, the first higher, placental mammals with their much-expanded brain cortices appear. It thus seems as if *steps of more marked lateralization of brain functions are paralleled by steps toward the separation of the main continental blocks of the Earth.*

In the old paradigm it would be natural to ask: Why would these processes occur in parallel? In that paradigm the two processes, biological evolution and plate tectonics, were regarded as essentially unrelated and unplanned. Yet as in the case of gender differences, the separation of brain functions and of continental blocks are both correlated with the beginnings of Days. This provides another example of how Halos on different levels of the universe, in this case those of the Earth and of the animals, are entangled and thus synchronistically affected by the changing quantum states that the Mammalian Underworld gives rise to. The two phenomena are hardly related in a physically causal way, where one causes the other. Rather in our nonlocal universe they progress in synchrony because the Halos of the two levels are entangled. As we saw in the previous section (figure 6.6), these polarizations on the level of the Earth and the biological organisms seem to be reflections of parallel effects on the galactic level. This is further evidence that different levels of the universe (figure 1.5) are entangled and thus follow the same alautun rhythm for the activation and deactivation of yin/yang polarities (figure 6.2). Such a synchronization of processes operating at different levels may serve in an ideal way to generate ecosystems to which the newly emerging animals are adapted, and the formation of continents obviously is the most basic aspect of these ecosystems and their climates.

Yet we must ask why plate tectonics may be related to the introduction of yin/yang polarities according to the rhythm of the Mayan calendar. To find a hint to the answer, we may note that in the mythologies of many of the ancient peoples of the Earth, the so-called World Mountain played a prominent role. Sometimes we see this World Mountain with a world axis, or Tree of Life, emanating from it, as in the Buddhist stupas. Some suggest that the many pyramids independently built by so many cultures were meant to honor the World Mountain, which was thought to harbor the intelligence that created the world. The Mayan pyramids for instance were built to reflect how the Earth was created by this World Mountain according to the rhythm of its particular Underworlds. The word *Underworld* thus takes on a more realistic meaning if it had an origin in the center of the Earth and the World Mountain that was located there. I suspect that the inner core of the Earth plays a much larger role in the evolution of life on Earth and holds much more information in its crystalline structure than what is currently known.

Figure 6.8 shows a cross section of the interior structure of the Earth, with the Earth's crust and its continents "floating" on mantle. This mantle envelops the outer core of the Earth, which envelops the inner core, which in turn envelops the recently discovered inner inner core. The movements of the continents on the surface of the Earth is believed to be caused by heat generated by the radioactivity in the interior of the Earth, a heat that creates plumes of convection streams in the mantle (similar to the way a convection stream goes upward in heated water). Given this understanding, it is logical that if there were a pattern in the continental drift, this would ultimately go back to periodic changes in the inner crystalline core of the Earth brought about by the introduction of yin/yang polarities. How exactly this would come about

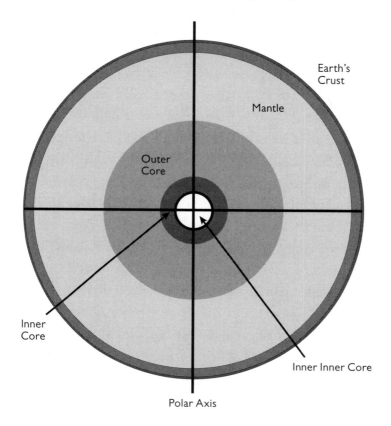

Figure 6.8. Planetary Tree of Life (polar axis) and its relationship to the interior structure of the Earth. The interior of the Earth is divided into layers with different solidities, temperatures, and chemical compositions. While the crust, mantle, outer core, and inner core have been known for about seventy years, the inner inner core was discovered in 2008.

is mostly a matter of conjecture, given our inadequate knowledge about the interior of the Earth. Yet the inner core is known to have a symmetry line (meaning that it is elongated) along the polar axis.[15] This would thus be the resonance line for the Cosmic Tree of Life, from which its creative pulses in three dimensions would emanate. If this is true such shifts could create directed convection streams that lead to the observed changes in the continental structure of the Earth at alautun shifts. This is supported by the recent finding that convection streams are emerging under Africa and the Pacific Ocean, based on a quadripartite distribution of chemical perturbations in the lower mantle.[16]

Hypothetically then at the shift points between Days and Nights in the Mammalian Underworld, the yin/yang polarities, corresponding to the shifting quantum states of the Cosmic Tree of Life, influenced the Earth's inner core in such a way that this directed the convection streams that generate the continental drift on the Earth's crust. Of course, convection streams generated from the interior of the Earth would not create perfectly straight coastlines for the main continental blocks, but they would be at the root of the rough separation for instance between Eurasia and the Americas. According to this hypothesis the entire geological evolution of the Earth, and the way it has become a hospitable planet with a stable temperature and continental structure surrounded by seas, ultimately goes back to the effects that yin/yang polarities have had on its inner core. This hypothesis explains why the separation of continental blocks parallels the emergence of lateralized brains, as the two would be affected by yin/yang-polarity shifts in a synchronized way. Viewing the inner core of the Earth, the World Mountain, as a source of life would then make sense.

The yin/yang polarities introduced by the Cosmic Tree of Life through the midplane of the galaxy would then synchronistically be introduced through the midplane of the Earth, and through the brains of bilateral animals during the Days of the Mammalian Underworld (figure 6.6). The wave movement of Days and Nights, brought about by the shifting quantum states of the Cosmic Tree of Life, thus seems to be synchronistically reflected at least on three different levels: galactic, planetary, and organismic. An interesting conclusion from this is that *the process of biological evolution is not something that takes place on our own planet in isolation from the rest of the universe.* Instead, all the different levels of the universe are participating in synchronized entangled processes with the purpose of generating life. As pointed out before, biological evolution cannot be understood merely from the biological mechanisms for the emergence of species. Equally important for the emergence of life, on our own or on any other planet, are the entangled processes on a universal, galactic (including the Halos of "dark matter"), and planetary scale that create the physical context for life to emerge. In fact, the whole universe seems to be involved in such processes, giving rise to the larger ecosystems that are just as intelligently designed as those generating cellular or animal life.

In a sense, what is suggested here is a rational explanation for why the Earth shares many of the characteristics of a living organism, a theory known as the Gaia hypothesis, originally proposed in a scientific context by James Lovelock in 1970. His theory incidentally is one of the few based on a holistic approach that has gained some recognition in mainstream science, although there has always been a significant resistance to it because it is not really consistent with Darwinism. Gaia, the living planet, has for instance not evolved because of natural selection, but is a particular level in a hierarchy of organizations of life, including the galaxy and the cosmos as a whole, whose evolutionary processes are synchronized by the introduction of yin/yang polarities on different levels. Through the effects of its Halo, the current "living" organism of Gaia has its origin in plate tectonics, which generated homeostatic mechanisms for life-supporting physico-chemical processes on the surface of the Earth. These mechanisms, I suggest, find their origin in the inner core of the Earth and in the resonance of this with the Galactic and Cosmic Trees of Life, which shift polarities in accordance with the quantum states described by the Mayan calendar. The Gaia hypothesis is fully consistent with the Tree of Life theory and may in fact be seen as a logical consequence of this theory.

MASS EXTINCTIONS

The biological history of our planet is not only about creative pulses transforming its structure to support the generation of new species. Of almost equal importance is the extinction of species. Thus, the overwhelming majority of organisms that have inhabited the Earth have eventually become extinct. In the Mammalian Underworld millions of individual species became extinct at one time or another, and the average time that a species has lived on Earth is estimated at only a few million years. In addition to the ongoing emergence and disappearance of new species, there have also been mass extinctions, times when major portions of the species of our planet have gone extinct, never to return again. The four most important mass extinctions in the past 600 million years are summarized in table 6.2. Again, I have not made a biased selection, but simply included those, and only those, that were listed by the source. The extinctions summarized in this table were massive, were relatively sudden, and involved whole classes of organisms. (A fifth extinction that occurred 205 million years ago is also sometimes included among the big ones.) We will determine whether these can be understood from a model of evolution that is based on a nested hierarchy of entangled levels.

The causes for mass extinctions proposed within traditional research vary, but usually climate or atmospheric changes, asteroid impacts, competition with other species, changes in sea water levels, explosions of nearby supernovas, or a combination of these events have been suggested. It is commonly, but not always, the case that a single cause for an extinction cannot be identified. In table 6.2 however it is immediately apparent that the major mass extinctions occur close to shifts between Days and Nights in the Mammalian Underworld (extinctions in the Cellular Underworld were discussed in chapter 4). I would like to suggest that on a deeper level the causes for the mass extinctions are to be found in the combined effects—at different entangled levels—of the yin/yang-polarity shifts occurring at those times. What might then have caused these extinctions and how are they linked to the shifts in the corresponding quantum states of the Cosmic Tree of Life?

To understand this we need to look at how the causative factors might have been synchronized. There may be not only immediate causes, such as a change in sea level, but also deeper causes, such as a change in the inner core of the Earth, hypothetically brought about by

TABLE 6.2. THE MAJOR EXTINCTIONS IN THE MAMMALIAN UNDERWORLD

Heaven (Day or Night)	Beginning (Myr)	Major Extinction[a]
Day 1	820.3	
	757.2	
Day 2	694.1	
	631.0	
Day 3	567.9	
	504.8	
Day 4	441.7	Extinction of 85% of all species (brachiopods, nautiloids, and many trilobites, 440 MYA)
	378.6	Extinction of 82% of all species (primitive fish and coral reef invertebrates, 380 MYA)
Day 5	315.5	
	252.4	Extinction of 96% of all species (marine invertebrates, bivalve mollusks, all trilobites, and sea scorpions, 245 MYA)[b]
Day 6	189.3	
	126.2	
Day 7	63.1	Extinction of dinosaurs, pterosaurs, plesiosaurs, and ammonites (65 MYA)

[a]Philip Whitfield, *From So Simple a Beginning: The Book of Evolution* (New York: MacMillan, 1993).

[b]Most current sources would place this extinction closer to the actual beginning of Night 5 at about 251–252 million years ago (see page 190).

shifts in quantum states, which through plate tectonics may have caused the change in sea level to begin with. Given the role of plate tectonics in causing volcanism, changes in sea level, changes in climate, and atmospheric concentrations of gases, it is entirely possible that periodic transformations of the inner core of the Earth are at the root of most of these extinctions, and it is also easy to understand why singular causes for them may be difficult to identify.

The best way to see how synchronized yin/yang-polarity shifts on different levels would play out is to look at an example, and a good one is the Permian-Triassic extinction, an extinction for which many different causes have been suggested. This occurred at the boundary between the Permian and Triassic periods, about 250 million years ago* (during the Mammalian Underworld), and is believed to have brought the most extensive extinction of biological organisms of all time. It included the extinction of 96 percent of all sea-living and 70 percent of all land-living species.[17] The trilobites (see table 4.2), which had been on Earth for several hundred million years, went extinct at this time. While life in the sea remained quantitatively the overwhelming portion of the biosphere, the Permian-Triassic mass extinction effectively and irreversibly shifted biological evolution from sea to land. When it comes to explaining the causes of this extinction, the point to realize is that when a quantum shift is synchronistically manifested by Halos at several different levels, it will result in a significant amount of stress to the habitats of the biological organisms.

From the perspective of this book, the Permian-Triassic extinction was caused by a combination of factors originating at different entangled levels in the nested hierarchy of life. One immediate cause might have been the massive restructuring of the continents, as mentioned in the previous section, when these merged into the Pangaea continent, with a concomitant change in sea level and a severe climate

---

*According to the Mayan calendar, the decisive shift point for the beginning of the fifth Night would have been 252.4 million years ago. The current best estimate in the scientific literature of the Permian-Triassic extinction is 251.4 million years ago. S. A. Bowring, D. H. Erwin, Y. G. Jin, M. W. Martin, K. Davidek, and W. Wang, "U/Pb Zircon Geochronology and Tempo of the End-Permian Mass Extinction," *Science* 280 (1998): 1039–45, doi:10.1126/science.280.5366.1039.

change. Greater volcanism may have resulted from the plate tectonic activity. All of these changes hypothetically have their origin in a shift of the inner core of the Earth that occurred as the fifth Night of the Mammalian Underworld began. It is also possible that the concurrent transition from one galactic hemisphere to the other may have had a direct effect on life on Earth (figure 6.6).

Since shift points between Days and Nights affect not only the geology of the Earth, but also the emergence of new species (table 4.2), they may also as a secondary effect bring about a shift in the balance between different species, leading to mass extinctions of some of them. The emergence of a new type of plant on land, the gymnosperms, may have shifted the ecological balance, leading to new forms of cooperation and competition between species, favoring some and disfavoring others.

We should also note that the beginning of the most serious of all the mass extinctions, the Permian-Triassic, coincides with the beginning of the Night that is ruled by Tezcatlipoca, in Aztec cosmology the lord of darkness. In almost any Underworld this fifth Night is a time of serious destruction, and so from the point of view of the Mayan calendar, this mass extinction was entirely predictable. This does not mean however that mass extinctions only occur at the beginning of Nights. The beginning of Days, with the introductions of new yin/yang polarities, which typically lead to bursts of new animal species, also cause tension on several levels that results in significant geological and ecological shifts, and these may also lead to extinctions. I should also point out that the new theory of biological evolution does not deny the struggle for survival in nature. Predators are everywhere and earthly transformations create significant threats for many species and so natural selection clearly plays a role in extinctions. The point is however that this struggle for life plays no role in creating the wave forms of biological species. Quantum jumps to new species do not—and cannot—happen because of natural selection.

There may have been a more direct influence on biological organisms caused by the shift in the quantum state associated with a new Day or Night. Each new Day or Night generates new species that are better adapted to the new environment created by that period than those species that had been generated by earlier quantum states. It then seems logical that the new species that were brought into existence by the most recent Day or Night would outcompete the earlier species that are less fit for the corresponding state, and this may be part of the mechanism behind the mass extinctions. This would explain why the classes of species that go extinct in mass extinctions often have been around for a long time, whereas species that recently emerged may survive. We find a typical example of this in the Permian-Triassic extinction, which brought about the end of several classes that went back to Day 3, such as the trilobites, whereas the more recent fishes from Day 4 and the reptiles from Day 5 survived. The reason then that newly emerged classes of animals have survived during the course of evolution is that they are in better resonance with the most recently generated quantum state.

Since the shifts between Days and Nights happen in synchrony at the different Halo levels, it is in most cases impossible to identify any single cause for mass extinctions, and it is thus not very surprising that a combination of causes has often been suggested to explain them. It is actually a logical consequence of the nested hierarchy of levels of biological evolution that there be multiple causes of mass extinctions, causes that may never be sorted out. Only on the level of the Cosmic Tree of Life would there be a single cause for mass extinctions, namely shifts between quantum states.

The mass extinction that may be the most famous of all occurred 65 million years ago and led to the annihilation of the dinosaurs at the boundary between the Cretaceous and Tertiary periods, only 2 million years before the beginning of the seventh Day of the Mammalian Underworld. Most researchers agree that this was caused by the meteorite that created the large Chicxulub crater on the Yucatán Peninsula in southern Mexico.[18] This impact also caused atmospheric changes that led the skies to darken and the temperature to both rise and fall drastically over the following time period. It has been argued that if the meteor that created the Chicxulub crater had been only twice its actual size, it might have caused the extinction of all higher life on Earth. This meteor explanation seems well supported, since all across the planet an increase of the element iridium, which is typical of meteorites, has been found in sedimentary layers coinciding with this particular impact.

Even so, this mass extinction may be viewed from a larger perspective, which may reveal that it had a more complex background than is usually recognized. To begin with, we may notice that this extinction occurred at about the same time that the solar system passed through the midplane of the galaxy (figure 6.6) and was located almost at the center of a large-scale celestial event that gave rise to a ring of stars, including the formation of the Pleiades and the Belt of Orion, that goes by the name of Gould's Belt. It is possible that the meteor that hit Earth is related to this dramatic event of star formation and in this way corresponds with a yin/yang-polarity shift introduced on a galactic level by the seventh Day. At two other levels, the emergence of the higher mammals and the complete separation of the continental blocks of the planet, we also find factors that might have contributed to the extinction.

Regardless, it needs to be explained why the dinosaurs—and not the mammals—went extinct at this time. While factors such as body size and temperature may have contributed to the demise of the dinosaurs, I would still like to suggest incompatibility with the new quantum state as a primary cause. The dinosaurs, which had small brains with little lateralization, may have been inconsistent with the new yin/yang polarity introduced at the beginning of Day 7. The resonance of the dinosaurs with the new quantum state of Day 7 was clearly inferior to that of the mammals, since these had much-developed forebrains, separate brain functions between the left and the right halves, and substantially expanded cortices. The higher mammals were in fact a product of this very new yin/yang polarity, and their brains were hence much better adapted to this than those of the dinosaurs. Or, put in other terms, the lateralized brains of the mammals were aligned with the yin/yang-polarity field introduced by the Cosmic Tree of Life at the beginning of the seventh Day of the Mammalian Underworld. Thus they had the potential to continue to evolve with the Anthropoid and Human Underworlds to a higher level of intelligence. Because of such mechanisms, where the brains of organisms must be adapted to the yin/yang polarity ruling at any given time, the evolutionary plan seems to find its way forward in one way or another.

In the next chapter, as we look at how biological organisms are created "from below," we will discuss how changes in wave forms of species, brought about by shifts between Days and Nights, may have a direct effect on both the emergence and extinction of species. At this point suffice it to say that no human being has ever experienced anything even close to the powerful quantum shifts that occurred in the Mammalian Underworld. Human beings emerged into an already existing, for us quite comfortable, yin/yang polarity when the dramatic transformations of the continental structure and ecological systems in that Underworld were long gone. That we ourselves have not experienced any similarly strong yin/yang-polarity shift should however not be a reason for us to deny that they may have had a powerful effect in the past.

When it comes to extinctions, it seems clear that biological evolution cannot be looked upon in isolation from the universe at large and that the processes leading to the emergence and extinction of species on different levels of the universe are related. These processes at different levels are connected because their Trees of Life are entangled and subordinated to the same overall purpose of the evolution of an Underworld. Since mass extinctions (and incidentally also the more common extinctions of single biological species) are part of a designed evolution, it is necessary to look upon them in a way that is different from what most Darwinist thinkers do. They typically emphasize only the negative aspect of mass extinctions, but I think it is important to note that extinctions also create the space for something new to emerge. This was clearly the result of the extinction of the dinosaurs, whose brains were not as lateralized as those of the mammals and hence would not have been able to fulfill the later mental evolution that is part of the climb up the Cosmic Pyramid. A continued dominance by the dinosaurs would most likely have blocked the evolution of the mammals into anthropoids and humans. A mass extinction is often more of a transformation of the galactic and planetary context for life than a destruction of this context, and so *the emergence of new species is inherent in the same process as the extinction of old ones*. Both the emergence and extinction of species are part of the cosmic plan, and we may recall that Shiva, the Hindu god behind the 108 transformations of creation, was both a creator and a destroyer—destroying in order to create the space for something new to emerge—and this may serve as a good mythological allegory for mass extinctions.

COMMON DESCENT?

We have now seen how not only the emergence of species, but also their extinction is a quantized phenomenon in which the crucial events occur close to some of the most important shift points in the Mayan calendar. For this reason it is meaningful to discuss the relationship of the quantum phenomena of the Cosmic Tree of Life to speciation in more detail. The higher mammals emerged 63.1 million years ago, when twelve hablatuns of the Cellular Underworld and twelve alautuns of the Mammalian Underworld had passed. This corresponds to a new quantum state of the Cosmic Tree of Life, and to entangled lower levels. Hence, a species, or rather its wave form, that emerged at that particular time could be assigned the quantum number 12.12.

The Mayan calendar can be subdivided into kinchiltuns, kalabtuns, pictuns, and so on in the tun-based system, and each of these time periods contributes aspects of the wave forms, or Platonic archetypes, in three dimensions from which the different species are created. Thus if a species emerged 59.9 million years ago, it would be defined by the quantum number 12.12.1. The added number 1 then signifies that a kinchiltun of 3.2 million years had passed since the beginning of the seventh Day of the Mammalian Underworld, something that reflects a new quantum state on the level of the Cosmic Tree of Life. The beginning of a new kinchiltun means a lesser change in the wave form of the species, compared with the beginning of a new alautun, because the longer time periods correspond to greater quantum shifts.

While the quantum number 12.12 theoretically could contribute only 144 different quantized wave forms ($12 \times 12 = 144$), each potentially generating a special kind of species, the addition of kinchiltun shifts would generate 1,728 of these wave forms ($12 \times 12 \times 12 = 1,728$). If we include all the quantum numbers at even smaller shift points down to the tun, as many as 430 billion different wave forms could be generated. This is far more than the number of species on our planet, which is roughly 10 million.[19] Thus while the change in wave form contributed by the kinchiltun, or shorter periods in the tun system, is expected to be smaller than those generated by the hablatun or alautun, the changes in quantum states at all of these time periods add considerably to the potential for biological variation. The biological variation in the species corresponding to the different quantum states is expressed geographically through the different polarizations of the Cosmic Tree of Life shown in the Cosmic Round of Light (see figure 6.3). This means that the emergence of new species will also be a function of how different species evolve in different locations, including the different continents, relative to the geographical directions.

Even if the shifts between Nights and Days in the Mammalian Underworld (corresponding to the most significant shifts between quantum states of the Cosmic Tree of Life) may be the most important points in time, when new multicellular species leap into existence, new distinct species emerge at many other times as well. The evolution of a species may therefore be looked upon as a result of a sequence of quantum shifts in the Cosmic Tree of Life, and an enormous biological variation is in principle possible in the lineages created by these shifts.

(When realizing this we may remind ourselves of the ancient myths saying that the Tree of Life is the origin of all life.) The evolution of a species through a sequence of minor shifts could give the impression of a gradual change, as indicated in the fossil record of horses.[20] At other instances the change may be sudden, such as when a major quantum shift took place as a new hablatun began 1.26 billion years ago, resulting in the emergence of eukaryotic cells. The model presented here may thus explain how evolution can occur both as slow, gradual change and as sudden leaps, depending on the magnitude of the change in quantum number.

New species, in other words, emerge at "cycle shifts" in the Mayan calendar precisely because it is at those particular times that a new combination of polarizations are introduced. We have seen that the different Underworlds play different roles in crafting the three-dimensional body plans of animals. At every onset of a new Underworld, a "quantum jump" takes place, resulting in fundamentally new kinds of species, such as unicellular, bilateral, or erect. The emergence of any new species is brought about by a particular combination of influences, which results from the polarizations, or quantum states, of a particular time period in the Mayan calendar. We can therefore define the human being (which emerged when the thirteenth kalabtun in the Human Underworld began in the thirteenth kinchiltun in the Anthropoid Underworld in the thirteenth alautun in the Mammalian Underworld in the thirteenth hablatun of the Cellular Underworld) as a species that was created by the wave form with the quantum number 12.12.12.12. Just as there is no atom that is a mixture of an oxygen atom and a nitrogen atom, *each species is distinct and of a unique kind generated by a specific quantum number.* 12.12.12.12 may then for instance be regarded as the quantum number defining the state of polarization of the Cosmic Tree of Life 158,000 years ago, which is exactly the state that gave rise to the wave form of the human being.

It is very likely that many of the different species created by these quantized wave forms were competing in such a way that only some of the possibly 430 billion species would survive. Thus, the Darwinist mechanism of natural selection does play a role in the destruction of less fit species, but obviously plays no role in the creation of the wave forms from which these species emanated in the first place. Examples of survival of the fittest are provided in the mass extinctions discussed in the previous section. Such a reproductive competition, or struggle for survival, however does not mean that the species were created by natural selection or that the species emerged from a series of random, undirected changes in their genetic or epigenetic composition. Nor does it mean that these species were created in response to environmental factors. Rather, the different polarities induced by the Cosmic Tree of Life are quantized, and ultimately the physiology and biochemistry of these species are subordinated to the quantized wave forms from which they emanate. *Without such quantized wave forms there would be no biological species to begin with.* Because the emergence of species is quantized, there are persistent gaps, or missing links, in the fossils record that will never be filled. Very often species emerge through "quantum jumps," such as from a land-living mammal to a whale.

The process of biological evolution is fundamentally the result of shifts between quantum states, and every species is associated with a particular quantum number defined by the point in time when it originated. This quantum mechanism explains not only the persistent missing gaps between species in evolution but also the inability of human beings to create new species by inducing mutations. Humans are unable to create new species because these are not created by mutations, but by wave forms. In the same way an electron cloud of an atom is defined by a combination of quantized electron orbitals, the three-dimensional Platonic body of a species is defined by the combination of yin/yang polarizations in different Underworlds emanating from the Cosmic Tree of Life, and not by genes in the DNA that mutate.

The quantized nature of biological species may also explain why it seems never to be possible to state with certainty that one species has evolved from another. This is problematic for Darwinism since its theory postulates that linear evolution takes place everywhere in nature. Yet when asked whether the human being evolved from the chimpanzee, the Darwinist answer will be "No, but they had a common ancestor." If you ask if *Homo sapiens* evolved from the Neanderthals or some other hominid, the answer will almost certainly be the same. These answers are correct, but not the underlying theory, since in reality the emergence of new species is a quantized phenomenon. Even though Darwinism postulates descent with modification, no linear descent is ever strictly demonstrated in practice and instead, more often than not, reference is made to some hypothetical common ancestor that has never been found. Nevertheless, the data that speak against the linear descent of species cannot be ignored. The reason for the absence of evidence of such direct ancestral forms is not a lack of fossilized species, but that each species emerges in a distinct form as a result of a quantum shift in the Cosmic Tree of Life.

Biological evolution is thus quantized, and each species is unique. It emerges because of a quantum shift in the Cosmic Tree of Life, not because of a continuous linear change. Each quantum shift generates a new wave form from which new species may be generated, and each species has a distinct essence that is not shared with any other. We may talk about predecessors or classes of animals within which the wave forms of the species are more similar compared with others. Yet there is no descent with modification because of small, gradual changes, as Darwin thought. Hence, for instance, while the wave form, and corresponding quantum number, of the human being is more similar to that of a chimpanzee than to a crocodile, neither the human, nor the crocodile, nor the chimpanzee descended from any other species. There is no continuum between different species and no direct lineages exist between them. Hence, the phylogenetic tree shown in figure 3.2 does not really give a true picture of reality. Instead, each species starts a new line in accordance with the different quantum numbers that would correspond to the time scale on the Y axis.

Given our current understanding, we should not be surprised if a

Darwinist researcher described the genealogical tree of life of plants as a *blurry, inscrutable thicket.*[21] What Darwinism in other words calls the tree of life simply does not exist, and the real Tree of Life is in an entirely different place. This discussion pertains to our own planet, but if we consider life on other planets in the universe it is even more obvious that the idea of common descent lacks any basis.

IS ANYBODY OUT THERE?

Considering that biological organisms have now been found to be the result of a purposeful evolution operating on earthly, galactic, and universal levels, the long-standing question of whether there is extraterrestrial life may be viewed in a new light. Thus at least since the mid-eighteenth century, people have been speculating about the possible existence of life, especially intelligent life, on worlds other than our own. For a long time it was for instance held as a fact that there was life on Mars that had built what seemed like canals on its surface, although at the time any prospects for actual contact with such purported extraterrestrial civilizations then seemed very remote. In 1947, the UFO frenzy started in the United States when people reported seeing essentially galaxy-shaped objects in the sky that were assumed to have an extraterrestrial origin. This had been preceded by reports in Sweden the year before of inexplicable "flying cigars" in the night sky. As the space age began with the Soviet launching of *Sputnik* in 1957, interest in extraterrestrial life naturally increased even more, and the whole field of science fiction came to largely focus on fleets of intergalactic space ships. After the American Moon landings, films like *ET, Close Encounters of the Third Kind,* and *Star Wars,* as well as the *Star Trek* series on TV, captured the imagination of millions. If nothing else this has shown that there is a widely held belief that we are not the only intelligent life form in the universe.

A scientific approach to solving the riddle of extraterrestrial life was the establishment in the 1960s of SETI (Search for Extraterrestrial

Intelligence), which aimed to pick up radio signals with an intelligent origin from space and also attempted to send messages to distant star systems in the hope of receiving responses. As yet, nothing conclusive seems to have come out of these efforts, partly because of the inherent difficulty of using (in this context) slow-moving radio signals that would require four years to reach even our closest stellar neighbor. A number of scientific breakthroughs in the 1990s however seemed to create a better foundation for the search for extraterrestrial life. One was the much-anticipated discovery that indeed stars other than our own Sun have planets orbiting around them. By 2008, more than three hundred of these extrasolar planets had been discovered,[22] including rare planetary systems that could well be similar to our own.* Two other relevant events receiving attention in the 1990s were the discovery of bacteria on meteorites[23] and of extremophiles.[24] Extremophiles are unicellular organisms that not only survive, but also thrive, in harsh environments. This seemed to increase the prospect that simple life could survive under the more strenuous conditions that exist on other planets or moons in our solar system.

In the 2000s, however, there seems to have been somewhat of a reaction to these hopes, not only in the form of a decreased interest in the UFO phenomenon, but also because new scientific data seem to indicate that more advanced forms of life might be more unusual than previously thought. Although there may be billions of galaxies, each hosting billions of star systems, it has become emphatically clear that far from all would be hospitable for life. The vast majority of extrasolar planets have seemed very exotic and unlikely as stages for life. Typical among them are the hot Jupiter-sized planets revolving very close to their stars at high rates, which could hardly harbor life

---

*A solar system with both a Jupiter-like and a Saturn-like planet has recently been discovered. B. S. Gaudi et al., "Discovery of a Jupiter/Saturn Analog with Gravitational Microlensing," *Science* 319 (2008): 927–30. To identify other Earth-like planets, a new generation of Extremely Large Telescopes (ELT) is now being built with which the direct detection of Earth-mass planets is expected to be within reach.

because of their extreme temperatures. Other extrasolar planets have very elliptical orbits, which would also preclude life because of their extreme shifts in temperature. The idea has thus gained ground that our own Sun is a rather special star and maybe the conditions for life here are indeed more narrowly dependent on its metallicity than has previously been thought.

It has also increasingly become accepted in the scientific community that the universe is fine-tuned for life and our understanding of this fine-tuning in fact allows us to exclude many star systems as potential harborers of life. The existence of higher life on our own planet seems to have depended not only on the fine-tuning of universal constants, but also on a series of seemingly fortunate events, such as the formation of a moon at just the right distance to contribute to the exact declination of our polar axis, which is necessary for a stable climate, a number of meteor impacts, the establishment of plate tectonics, and so on. Based on such considerations it becomes more reasonable to believe that the emergence of higher life may be a relatively rare event in the universe. Yet our estimates of the likelihood of life on other planets will depend very directly on what model we are using to explain the emergence of life. It seems obvious that the likelihood of finding life elsewhere in the universe will very directly depend on whether you regard life as the result of an intentionally designed plan or if you regard it as an oddity that was not really meant to be. Equally profoundly, it will depend on whether you look upon the emergence of life on our own planet as a result of processes operating on a cosmic and galactic scale or merely of coincidences created locally here in isolation from the rest of the universe.

The Drake equation is a commonly used procedure for estimating how many planets with life there may be in the galaxy with which we could communicate. This equation is a simple multiplication of several fractions, f_i, of the total number of stars with planets in the galaxy, N, that meet certain conditions that are believed to be necessary for such communication. (If we were interested in the total number of such planets in the universe we would also have to include an estimate of the fraction of galaxies in the universe that could harbor life.) The Drake equation was used as part of the SETI effort and included a number of conditions, such as the fraction of stars that had planetary systems, f_p; the number of planets per such solar system that would be able to harbor life, n_e; the fraction of such planets where intelligent life evolves, f_I; the fraction of planets with intelligent life that communicates, and f_L; the fraction of a planet's lifetime during which such communicating civilizations may be active. By multiplying these fractions, an estimate of the number of planets with life that might be able to communicate with us, N_{life}, is thus attained:

$$N_{life} = N \times f_p \times n_e \times f_I \times f_L$$

This equation was never meant for anything other than illustrative purposes, since there has been so little hard data available to estimate the values of the various factors. The various guesstimates that have been made with it have primarily been reflections of the optimism or pessimism of the individuals making them. However, with the discovery of exoplanets we have gained a somewhat better grasp of the situation. It has become clear that f_p, the fraction of stars with planetary systems, probably is very high, but that n_e, the fraction of planets where intelligent life could evolve, may be very low because of the narrow conditions required for this to happen. Two interesting books, the previously mentioned *Rare Earth* and *The Privileged Planet,* by Guillermo Gonzalez and Jay Richards, modify the Drake equation based on such considerations, and both, with somewhat different approaches, come to the conclusion that higher intelligent life may be very rare in our galaxy. The authors of *Rare Earth* conclude that while unicellular life may be very common, multicellular life is not, partly because it would be much more sensitive to factors creating mass extinctions. The authors of *Privileged Planet* go even further and formulate a modified Drake equation with twenty-two factors that would determine our ability to discover advanced life elsewhere. They conclude that most of these probabilities are fairly low, and they estimate that there would be less than one planet in the galaxy that could harbor life that we would be able to communicate with (>0.01 planets out of the 200 billion star systems).[25] They conclude that we are living on a planet that indeed is privileged.

The new theory of biological evolution presented here, however, changes the perspective on the potential existence of civilizations that we would be able to communicate with. First, the emergence of life can really no longer be seen as the result of the multiplication of a number of fractions, sometimes expressing probabilities of catastrophes such as meteor impacts that are looked upon as random factors with a given probability. In a universe in which the generation of life is a purposeful process, originating in overlapping morphogenetic fields, and not taking place independently in each solar system in isolation, it is necessary to look at the emergence of life differently. In order to identify planets with life it is necessary to look for stars that in metallicity and size are similar to our own Sun as well as for those that harbor planets with a tun orbit. But with regard to all the other factors and events that may contribute to life on such planets, it does not seem to make sense to look at each one of them, such as the existence of a moon of the right size and distance, or the right inclination of the planetary axis, as having a probability that is independent of the others. Instead, in the same way we could see that the galactic Halo contributes to the emergence of life in ways we cannot fully understand through its "dark matter," we may expect that the heliospheric or planetary Halos contribute to the emergence of life in *certain* star systems in ways that we may not fully understand. While from our current understanding it may seem like an accident with a certain probability that our planet was hit at the beginning of the seventh Day of the Mammalian Underworld by a meteor of exactly the right size to kill all dinosaurs and let the mammals survive (or that a planetoid collided with Earth at an early point to give rise to a moon at the right distance to stabilize the inclination of our polar axis, thus giving the Earth a stable climate), such an event may in fact have depended on the existence of a heliospheric Halo. Thus in the present model, certain solar systems may simply be tagged for the emergence of life because they are in an optimal resonance with the Cosmic Tree of Life. In such star systems all the conditions for life may be met, whereas in others none of them will be met. At the present point, suggestions concerning the mechanism for this would be speculative, but, again, we have already seen that the galactic Halo plays such a role for the emergence of life by modifying Newtonian mechanics, and so why would a heliospheric Halo not play a similarly life-supporting role?

If this is true the number of star systems with life will not necessarily be as low as was estimated by the abovementioned authors. Rather, life would emerge in star systems whose resonance with the Cosmic Tree of Life is optimal and not at all in others. It would then seem that it would not be possible to estimate the number of star systems with life by means of the Drake equation. The number of planets with life would instead have a "predetermined" value rather than be the result of a number of independent probabilities. Generally speaking, by seeing the new theory of biological evolution as a universal process, we might conclude that there is a much greater certainty that higher life does exist in other star systems, even though we do not know how common this may be.

I propose that rather than the traditional material factors included in the Drake equation, we should explore how the tun system can be used as a basis for the search of extraterrestrial life. To do so I believe that exobiology would have to become more "Mayan" and explore how the large-scale cosmic and galactic processes are related to the basic periods of this calendrical system. It is possible that we would then find that there are a considerable number of planets in the galaxy that harbor life. If this reasoning is true, the Milky Way is, to use an ancient concept, like a Cosmic Egg that may give birth to a litter of limited-size planets with civilizations that as yet are unaware of each other and probably are separated by large distances.

When it comes to the possibility of communicating with these other civilizations by means of radio signals, the Tree of Life theory provides a new view. According to this theory, life evolves with the same rhythm everywhere in the universe, since the evolutionary plan that began at the big bang emanates from the Cosmic Tree of Life. Unlike relativity theory, which does not recognize a preferred system of reference, the Tree of Life theory recognizes an absolute time emanating from the Cosmic Tree of Life, which is independent of any measurements that may be made of it by human observers. (This is not Chronos time, measurable by quantitative comparisons with physical cycles, but Kairos time, which is independent of the presence of observers.) In this absolute creation time there is a universal "now" defined by the Underworld and Heaven that rules in a certain location, and this "now" is the same everywhere in the universe regardless of whether time is measured there or not. Thus not only biological, but also mental, and hence technological, evolution would proceed in parallel everywhere in the universe.

As I have discussed in earlier books, and we can see in figure 5.4, the human perception of reality is under the influence of the time plan generated by the Cosmic Tree of Life, and this shifts between the different Underworlds. For example, take the field of electromagnetism, which we have only been aware of since the Planetary Underworld started in 1755. (Some phenomena of static electricity might have been noted earlier, but the systematic study of electromagnetism started around the mid-eighteenth century.) It was also within this Underworld, and not any earlier, that radio signals first came into use, about 100 years ago. If biological and mental evolution are synchronized all over the universe by the Cosmic Tree of Life, this would mean that extraterrestrial civilizations would have been capable of sending radio signals that we would be able to detect only for the past 100 years or so. Thus if an extraterrestrial civilization is able to attain the technological level that would allow it to communicate with radio signals, this would take place at the same time that we would be able to do so ourselves, and not at any earlier time. This limits the window for the emergence of communicable civilizations to approximately the last 100 years. Consequently, such communication by means of radio signals would be possible within a radius of only 100, or at the most 250, light-years from us. If higher intelligent life is as rare in the galaxy as studies of exoplanets suggest, we would have the greatest chance of communicating with extraterrestrial civilizations, at least via radio signals, by looking for planets with a tun orbit that are located on star systems within such a radius of ourselves.

7

The Biochemical Basis of Biological Evolution

A NEW DEFINITION OF LIFE AS A BASIS FOR UNDERSTANDING ITS ORIGIN

We have in the previous chapters looked at biological evolution from "above," that is at the emergence of life in the larger context of the cosmos, galaxy, and planet. From this study, perhaps the most profound realization is that it has become clear that life is not something that just accidentally happened to "pop up" at our own particular planet. The universe in its entirety is designed for the creation of life, and to bring about a universal process of evolution, the cosmos utilizes the lenses of a nested hierarchy of levels of organization. Because these lower-level Trees of Life locally reproduce the Cosmic Tree of Life, life on our own planet emerged in a specific context of space and time, where our own galaxy, heliosphere, and Earth were "tagged" and prepared for the generation of higher life. In this perspective, biological evolution can no longer be looked upon in isolation from the context provided by the cosmos at large. A more broadly shared insight that there is an overall mechanism for the evolution of life in the universe, brought about by its Central Axis, would amount to an enormous paradigm shift not only in biology but also in science in general and in society at large. It is a shift that provides for a completely different view of our own purpose and place in the cosmic evolution scheme, where we can see that we are cosmic beings in a much more literal sense than most people might think. Life, wherever in cosmos it may arise, is the result of a purposeful time plan operating at several different synchronized levels of organization. The synchronization of these levels seems to depend on their entanglement with the Cosmic Tree of Life, which retains seniority in the creation process. While it is hard to assess the limits to such entanglement, or teleportation of quantum states, it is very significant that it does occur between the different levels of the cosmos.

We are now faced with the question of the origin of life, and I think it is fair to say that conventional science has not had any success in addressing this whatsoever. The dominating hypothesis of conventional science is that life emerged from random reactions in some prebiotic soup, and such simulations are mentioned in biological textbooks as indicative of how life begins. Actual attempts to generate biomolecules from random chemical reactions however have not even come close to generating functional biomolecules such as DNA and even less so living cells. And how could it possibly be successful with such an approach? It is obviously impossible that meaningful information for the metabolism of a cell could ever be generated from random chemical reactions. The taboo in conventional science against recognizing any higher guiding principle for the evolution of life means that there is no choice but to assume that life is an accident, and hence its explanatory power suffers.

In this chapter we will look at biological life from "below" and study the cells and their constituents from the more common perspective of biochemistry and cell biology. We will initially discuss the definition of life and then go on to the origin of unicellular life and multicellular life. Through the recognition of the Cosmic Tree of Life we have however already come considerably closer to understanding the origin of life than previously. We have seen that life is a function of Platonic wave forms emitted by the quantized Cosmic Tree of Life, and we have seen how a chain can be tracked from the galactic to the planetary down to the organismic level, where these wave forms are focused. It is from such "Platonic" wave forms of a quantum nature that biological organisms may materialize. We have yet to explain how this materialization takes place, and to do so we need to understand how these wave forms interact with matter—the atoms and molecules from which biological organisms are formed. How do the wave forms interface with these atoms to create the biochemical processes of life? What wave form or forms are involved in creating unicellular life? What is the nature of the atoms that make up the molecules of life, and how did they emerge at the right time in evolution for the wave form to manifest as biological life? How were cells created?

Before answering these questions let us first consider what *is* this biological life that evolves. It then seems relevant to define what life is, since this is fundamental to how we understand its origin and evolution. The definition of life that will be used here is: *A living system is a system of molecules created by a wave form emanating from the Cosmic Tree of Life that maintains an energetic imbalance with its environment.* This definition implies that life is always *cellular* in nature. It would not include viruses or dead cells, since these do not maintain an imbalance with the environment. In contrast to many other definitions, this one does not include a requirement that the living system be able to reproduce itself. Reproductive ability seems more like a property of many living systems than a definition of what life is. An aspect of this definition that is not so common is the requirement for the system to maintain an energetic *imbalance* with its environment. Here it is proposed that this open cellular system out of balance with its environment is created and maintained by cellular Halos generated by the Cosmic Tree of Life. Such a definition of life, we may note, is consistent with a view of the universe that is driven to evolve by its inherent imbalance. Thus if a system is not permanently out of balance with its environment, it is not living. The energetic imbalance of a cell against its surroundings then gives rise to, and ultimately drives, a complex system of coordinated biochemical reactions that is referred to as its metabolism. If the imbalance cannot be maintained, the metabolism comes to an end and the cell dies, and this is exactly what happens when a cellular membrane is destroyed. If the cell membrane is destroyed, and a balance with the environment established, the cell dies even if all the molecules of the cell are still there. Life is thus not "located in" or a direct function of these molecules.

With this definition of life, our approach to understanding its origin becomes very different from that of conventional science, which describes life merely as a function of its molecules. Our point of departure is instead in the quantized Platonic wave form reality that lies behind life. This implies emphasizing the geometric aspects in order to understand how biological form can emerge. After all, when it comes to the three-dimensional forms of the species, we find that biological evolution is clearly synchronized with the Mayan calendar.

In contrast to our approach, experiments within mainstream biology study how the basic molecules in cells, such as amino acids (from which proteins are built) and nucleotides (from which RNA and DNA are built), could have originated. (I use the term *mainstream biology* rather than *Neo-Darwinism*, because the origin of life is outside Darwin's field of study. The approaches, however, are based on the same randomist philosophy.) Such experiments have the underlying assumption that the biochemistry and metabolism of cellular life have emerged from a series of random chemical reactions and that only the physico-chemical reality has relevance. A classical experiment to study the possible original formation of such basic molecules of life on Earth was performed in 1953 by Urey and Miller.[1] For a period of a week these workers let electrical discharges pass through a sealed vessel containing heated water and an atmosphere of methane, ammonia, and hydrogen, which was meant to simulate the conditions of the early Earth. Following this treatment the contents in the flask were analyzed and found to contain most, but not all, of the amino acids that are the normal building blocks of proteins.

Fatty acids and sugars also formed under these conditions. In experiments performed later under somewhat different conditions, some nucleotides, the basic components of the nucleic acids DNA and RNA, were also formed.

While such experiments might have something to tell us about the mechanism according to which many simple organic molecules may have formed, I think it is fair to say that they have provided no insights regarding the origin of life. Thus the hopes of many scientists that such experiments would be able to explain the origin of life have largely faded, and most researchers realize that life cannot be recreated in this way.[2] While the experiments were indeed successful in creating some of the basic organic molecules necessary for the emergence of life, and the reactions showed that the geometry of the orbitals of the atoms of life are favorable to creating such molecules, it did not seem as if they were leading us toward recreating life.

From the point of view of explaining the origin of life, random chemical reactions have serious flaws. The first flaw may initially seem insignificant, but it is a detail with crucial importance for the geometry of life, and hence I will explain it in some detail. Molecules that are based on carbon chemistry may potentially exist in two opposite forms of handedness, the so-called L- and D-forms, which are mirror images of one another. The two forms are called optical isomers, since they reflect polarized light in different directions, D- to the right and L- to the left. In cellular proteins and DNA, only one of these forms, the L-form of amino acids and the D-form of nucleotides, is found. Consequently, in living systems the double DNA helix corkscrews only in a right-handed way (figures 7.1b and 7.2). Living systems seem to have a preference for this corkscrewed form and synthesize only the components that have the particular D-form of nucleotides that may generate it.

In contrast, when the same molecules are formed in random chemical reactions, such as in the experiments described above, no selection of one form of handedness takes place, and nucleotides or amino acids

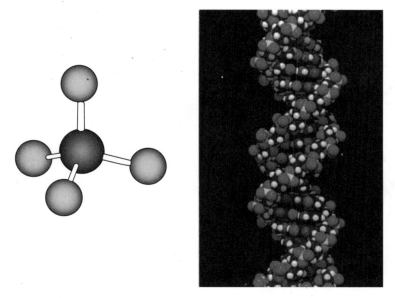

Figure 7.1. The golden ratio in the chemistry of life. Left: Methane molecule with an sp³-hybridized carbon atom at its center with an angle close to that of a pentagon (108 degrees). Right: In living organisms, DNA spirals (or corkscrews) in only one direction. The resulting form is called B-DNA, whose backbone is built exclusively from optically active D-nucleotides. With permission from Gerald Karp, Cell and Molecular Biology, *2nd ed. (New York: John Wiley and Sons, 1999).*

are formed in equal quantities of the D- and L-forms. Anyone who has worked with amino acid chemistry knows it is extremely difficult to separate the two forms by regular chemical methods. The selectivity in nature of just one of the optical forms seems to indicate that the large polymeric molecules found in nature, such as DNA, did not arise from components that had been formed by random chemical reactions. Not surprisingly maybe, most modern scientists believe that Earth life's "choice" of optical form was purely random, and that if carbon-based life forms exist elsewhere in the universe, their chemistry could theoretically be based on opposite optical forms. Curiously however, Louis Pasteur, who originally discovered this form of handedness of many organic molecules, had a different opinion. He concluded that any

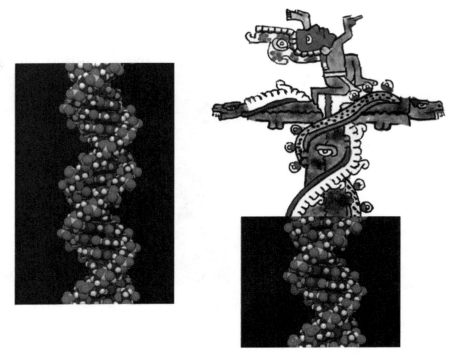

Figure 7.2. Left: Model of the DNA double helix. With permission from Gerald Karp, Cell and Molecular Biology, *2nd ed. (New York: John Wiley and Sons, 1999). Right: The DNA model combined with a detail of Codex Selden showing the corkscrewing into existence of a human being by the Tree of Life.*

hypothesis addressing the origin of life must also address the problem of handedness, and this is what we will do here.

The second flaw is that random reactions cannot have been the source of life, since by definition such reactions can give rise only to biomolecules devoid of information. This is in contrast to living cells, where DNA carries precise information that is expressed partly through RNA and proteins. Randomness does not generate meaningful information and could not possibly provide guidance for the kind of synchronized and coordinated system of metabolism that characterizes a living cell. The third flaw is that chemical reactions performed in the laboratory are unable to reproduce how a cell emerges as a distinct system out of balance with its environment. While it may be possible to seal off cells, and cell walls may form spontaneously if certain fatty acids are suspended in water, such vesicles are always in balance with the environment and hence dead.

THE CHEMISTRY OF LIFE

In chapter 5 we saw how the geometry of all chemical substances has an origin in the golden ratio. When it comes to the chemistry of life this becomes even more evident, because the golden ratio and the Cosmic Tree of Life play roles at an additional level and this can obviously not be the result of random chemical reactions. To see this we need to know that the golden ratio is related to the Fibonacci numbers, a series of numbers originally described by Leonardo of Pisa (1170–1250), who more famously goes by the name of Fibonacci. This series of numbers is formed by adding its two last numbers so that the next number becomes the sum of these: 1, 1, 2, 3, 5, 8, 13, 21, 34, 55, 89, 144, 233, and so on. Although the golden ratio was originally discovered in geometry, Johannes Kepler later discovered that the ratio between two consecutive Fibonacci numbers approaches phi as this series asymptotically approaches infinity: $55/34 = 1.6176$, $89/55 = 1.6182$, $144/89 = 1.6180$, and so forth.

The relationship between the golden ratio and the Fibonacci numbers may now give us a new perspective on the DNA molecule. In the established paradigm this molecule, which has been considered the primary factor for the expression of genes, is often seen as just another molecule that because of its structure happened to gain a special function. The idea has been that it is put together by consecutive random reactions. Yet we have already seen that it corkscrews in only one direction in living species and that it has an awesome double-helical structure that has no appearance of randomness. We may then ask whether there are any indications that this molecule instead has a direct origin in the Cosmic Tree of Life. This is all the more important to know considering that DNA seems to have a significant role in the biochemistry of individual cells as their proteins are generated by it.

Thus already the corkscrewing of the DNA, linked to the exclusive D-handedness of its components, indicates strongly that this molecule is a resonance unit for the Cosmic Tree of Life. Moreover, the double helix structure of the DNA is clearly designed based on phi and the Fibonacci numbers that phi is related to. The width of one round of the DNA double helix is 21 Ångström (Å), while its height is 34. The distance across its major groove is 21 Å, the minor groove is 13 Å, and a cross-sectional view of the DNA molecule from the top reveals two pentagons (whose angles are determined by the golden ratio). This means that the Fibonacci numbers 5, 13, 21, and 34 define the dimensions of the DNA helix.

The many expressions of the golden ratio in the geometry of the DNA ultimately go back to the angles between chemical bonds. Presumably because of the influence of the Cosmic Tree of Life on the orbitals of the Atomic Tree of Life, the atoms from which the geometry of life is built, carbon, oxygen, nitrogen, and to lesser extent sulphur and phosphorus (but no others), are what technically is called sp^3-hybridized. Thus instead of the perpendicular orbitals that were shown in figure 5.4, these atoms are at the center of tetrahedral configurations of orbitals where the angles between the bonding atoms are close to 108 degrees, which is typical of a pentagon (figure 7.1a). Since the pentagon is dominated by the geometry of phi, it can thus be said that the atoms, and hence also molecules, of life have the distinction of being created based on the golden ratio. These angles account for the properties not only of DNA, but of all other biomolecules, including notably the water molecule, which is also sp^3-hybridized. This geometry is what gives the water molecule the many remarkable properties that makes it crucial for biological life. Hence it is not entirely true when some scientists say that there is no fundamental difference between organic and inorganic matter. The difference resides in the geometry of the constituent atoms, where those forming the molecules of life are sp^3-hybridized. The presence of the golden ratio in the molecules of life is thus behind the fact that living organisms are formed by the kind of soft biological matter that they are.

The Fibonacci correlations support the argument that the DNA, the amazing double helix, is not something that emerged by accident from random chemical reactions in the early history of the Earth. Instead, it too was modeled upon the Cosmic Tree of Life and phi, and as such it is very likely to be ubiquitous and in fact an indispensable part of this universe. It can no longer be argued that the presence of DNA in all living organisms proves that they share a common descent from a single type of organism that happened to pop up on our planet. Rather, it seems likely that all life, everywhere in the universe, has a common origin in DNA generated in the image of the Cosmic Tree of Life.

CORKSCREWING LIFE INTO EXISTENCE

We have come a long way in understanding the origin of life and what generated the first single cells. To continue I feel we should again allow ourselves to be guided by ancient myth and simply look at a model of the DNA molecule together with a detail from the Mayan Tree of Life in the Codex Selden (figure 7.2). This ancient picture indicates that life, in this case exemplified by a human being that comes out of the Mayan Tree of Life, is corkscrewed into existence by a wave form (symbolized by the serpents) emerging from this Tree. Since we saw in the previous chapter that the Cosmic Tree of Life indeed is able to create systems in its image at lower levels down to the organismic, we may then assume that this may also be true when it comes to the emergence of cellular life.

I will here propose that the wave form that corkscrews unicellular life into existence is that of the DNA. Apart from the general biochemical plausibility of this and the stunning similarity of the DNA to the Tree of Life of ancient myth, there are also a few direct reasons to assume that it is created in the image of the Cosmic Tree of Life. The first is the observation mentioned in chapter 1, that polarized light from quasars seems to corkscrew around the Central Axis in the core of the universe. The second reason is the one just mentioned, that DNA corkscrews in only one direction (hence uses only the D-form of the potential building blocks, which rotate polarized light in a specific direction). These parallels are obviously consistent with the suggestion that DNA is the primary wave form emanating from the Cosmic Tree of Life. Moreover, the fact that both observations, although referring to phenomena at immensely different scales, pertain to the ability to rotate polarized light, seems to indicate that there is a deeper and more direct connection between the two levels.

What is suggested here then is that DNA does not emerge as the result of a series of random reactions that proceed step by step until a long helix emerges. Instead, it receives its information as a whole by resonance with the Cosmic Tree of Life, and hence it is created in the image of this *as a whole*. It is only the double helix that is created in the image of the Cosmic Tree of Life and not its building blocks. Thus DNA emerges as a whole, corkscrewed into existence from its resonance with the Cosmic Tree of Life, a possibility that has never been considered by mainstream science. Nevertheless, this hypothesis does have the virtue that it meets Pasteur's requirement that a theory about the origin of life must be able to account for the optical isomery of its cellular constituents. Hence the Cosmic Tree of Life can corkscrew in only one direction, meaning that the DNA in resonance with it can make use of only one of the two forms of optically active building blocks.

DNA, it should be noted, is a heavily negatively charged molecule. This means that if a Halo is generated around it, the Halo could hypothetically create a cell membrane or cell wall around itself. If this were to happen, the interior of the cell would be negatively charged, as cells are, and out of balance with its environment, a condition that as we stated above is required for the emergence of life. The DNA could then go on to create the RNA and proteins that it would need to sustain its metabolism. To account for how such a metabolism is formed, we will infer that DNA has a way of "seeing" its environment, analogous to the curious "eye" in the Tree of Life in the ancient picture (figure 7.2). This will be discussed in the next section.

At first sight these phenomena may seem to be outside ordinary chemistry, but as we pointed out above, the randomism of modern science is incapable of producing any alternative explanation to the origin of life whatsoever. Moreover, given that we have seen that the radii of atoms from which molecules like DNA are formed, as well as DNA itself, have geometries determined by the golden ratio, it is not difficult to accept that the double helix came into existence through such a mechanism of resonance. Since we have indeed found that higher forms of life are generated in the image of the Cosmic Tree of Life, why would we assume that lower forms of life, such as unicellular organisms, have a different origin? Similarly, since we have seen that consciousness in higher species emerges from their relationship of these to the Cosmic Tree of Life, it really is not so far-fetched to assume that DNA, created in the image of the Cosmic Tree of Life, has an ability to "see."

Why then, if life emerges from a resonance with the all-pervasive Cosmic Tree of Life, are cells not popping up everywhere around us? I believe the answer to this lies in the fact that the creation field of the universe is now very different from what it was in the Cellular Underworld. Thus the Cosmic Tree of Life is now in another quantum state. A chief factor here may be the entirely different rhythm of evolution that dominated an Underworld that was developed by creative pulses of a hablatun duration, 1.26 billion years long. Hence I believe that the processes of cell formation, corkscrewed into existence by the resonance of DNA with the Cosmic Tree of Life, could potentially take thousands if not millions of years in Chronos time. Before the onset of the Mammalian Underworld (or at least the pre-Mammalian Underworld, starting 1,640 million years ago), with its introduction of sexual polarity generating bifurcations leading to cell division, it is debatable if cells even divided, and if they did they might have done so at an extremely slow rate. I believe that the original emergence of cells through resonance with the Cosmic Tree of Life might be such a slow process that it could never be observed within the time frame of a human lifetime. Moreover, it is not something that could be reproduced in the laboratory, since human beings are unable to manipulate the Cosmic Tree of Life or bring it to speed up the process in such a way that we would be able to observe it. In our own day and time, most microorganisms such as bacteria replicate themselves at a very fast rate, and so it is easy to take for granted that this was always the case. Yet there are spores and microbes that may survive for very long periods of time without replicating, and I suggest that the same was the case for the early cells of our planet.

If we accept that the orbitals of the atoms (figure 5.4) were created in the image of the Cosmic Tree of Life, then it is not so difficult to accept that a large molecule like DNA could be formed in its image over considerable time periods. One of the things that I learned, the hard way, from my many years in an organic chemistry laboratory is that in chemistry everything happens. What happens depends essentially on how long time a reaction is given to occur. For all that we know, one turn of the corkscrew, amounting to ten base pairs, may have taken as much as a tun before the DNA had created proteins to help speed up the process. That the DNA molecule is thermodynamically stable, and a favored product in our universe, is evinced by the simple fact that it is ubiquitous in cells.

In addition to the geometry and chemistry of DNA, there is also evidence regarding its information content. Jean-Claude Perez, who has contributed a large and significant body of work, demonstrated in his books *L'ADN décrypté* and *Codex biogenesis: 13 codes et harmonies de l'ADN* (the latter published in 2009) that some of the numbers characterizing the genome at large reflect resonances with exactly those numbers that we have found to be associated with the Cosmic Tree of Life, such as 2 (for duality), 13, and most importantly pi and phi, the golden ratio. His studies show that the information carried by the DNA could not possibly have emerged as a result of random reactions or rearrangements of its component nucleotides, but that the entire genomes are constrained by these constants associated with the Cosmic Tree of Life. Thus the information carried by the DNA can only be looked upon as a whole. Depending on the organism studied, the constants defining the organization of the genome will vary somewhat, but it is noteworthy that the human genome, and that of the practically identical chimpanzee, are the only ones that in their entirety are defined by the constant phi. (The finding that the DNA sequence is constrained by the golden ratio has incidentally also been independently verified by other researchers.[3]) As we will see in the next chapter, this predominance of the Golden Ratio in the human being is consistent with the view that the human being is the end result of biological evolution. Perez argues that the central dogma of biochemistry clouds our understanding of the metainformation defining the amino acid sequences in proteins. He also maintains that one of the roles of the noncoding so-called junk DNA is to harmonize the genome with the constants of 13, pi, and phi. From our perspective here it is indeed very reasonable that an adherence to sacred mathematics may be crucial for harmonious resonance with the Cosmic Tree of Life. If this is true these mathematical constraints would actually protect against the kind of random mutations that Darwinists believe are behind biological evolution, and this is consistent with what was said earlier about the hazards to life of random mutations.

THE CONSCIOUS SEEING BY HALOS AND THE METABOLISM OF CELLS

A curious and yet very important aspect of the ancient Tree of Life is shown in figure 7.2 (right). We see an eye in its midst, and a pair of eyes can also be seen in the Mayan Tree of Life in figure 1.3. This would imply that the Cosmic Tree of Life, and by implication all organisms created in its image, is conscious, since seeing implies consciousness. This raises the question of whether the DNA molecule and its Halo, which emerged in its image from wave forms as life first originated, can be considered conscious. We might ask if there is any indication that the cells emerging from these wave forms were also able to "see" the reality they were surrounded by. This question is important because for a long time the development of cellular metabolism has been looked upon as a mechanical process that can be explained by reactions of purely inanimate chemical substances. If it can be reasonably proposed that the DNA is associated with a level of consciousness and cellular intelligence, it may be assumed that many aspects of cellular metabolism, including the nature of the proteins facilitating it, were created by the DNA over long periods of time in the Cellular Underworld.

In the theory presented here, everything that is created in the image of the Cosmic Tree of Life is conscious, which in practice means that everything that exists is conscious. Yet there is a tremendous difference in the level of consciousness of a human being and an electron. Since we saw earlier that consciousness evolved to a higher level with every Underworld, it can be said that consciousness emerges in response to the quantum state of the Cosmic Tree of Life. The consciousness of electrons, which came into existence immediately after the big bang, would then correspond to 0.0.0.0, whereas the consciousness of human beings would start at the quantum state of 12.12.12.12. The Cosmic Tree of Life can be looked upon as a giant quantum computer teleporting quantum states, but all interpretations of quantum theory in fact imply consciousness. Hence the Cosmic Tree of Life is a generator not only of life, but also of consciousness.

The DNA molecule can be said to be somewhere in between the electron and the human, but inasmuch as a Halo emerges around this molecule, so that a separate system of a cell is created, this consciousness will be considerably higher than that of an electron. In fact, I feel that the intelligence of DNA may have been strongly underestimated as we have not previously been aware that it originates in the Cosmic Tree of Life or that the sequence of DNA emerges as a whole and not as randomly assembled pieces. Could it be that the metabolism of the first cells was developed by such a DNA consciousness? The best way of identifying and localizing cellular consciousness may be to say that it arises from the interaction between the Halo and the DNA, which as we have seen serves as a molecular Tree of Life in single cells. With consciousness comes not only an ability to see, but also more generally to sense what is present in the environment and react to this.

Clearly, when higher unicellular cells, such as amoebas, see light or food they choose to move in the direction of these stimuli, which attests to a certain level of consciousness. In the bacteria that exist today, and even more so in higher eukaryotic cells, these sensory functions are however usually not attributed to the Halos, but to membrane proteins. Membrane proteins not only safeguard the imbalance of the cellular system versus the environment by controlling ion flow, but also serve as receptors for light, food, and many other kinds of stimuli. In response to such stimuli they signal to the DNA that it needs to adapt its metabolism to the circumstances. Our view of such processes is completely altered if we accept that they are created by the intelligence emanating from the Cosmic Tree of Life. In my view, these membrane proteins were created by the intelligence of the Halos and the DNA of these cells and were developed to take over the sensory functions originally assigned to these Halos because they were more effective at those tasks.

Hence, when discussing the origin of life and the emergence of the first cells, consciousness cannot be ignored. In a universe created by the Cosmic Tree of Life, consciousness is an inextricable part whether we like it or not, and in the DNA we have information directly emanating from this Tree. In this universe, nothing is completely inanimate. Thus I feel that the organization of biomolecules in a cell can never be understood merely from chemical reactions of what seem like inanimate molecules. Consciousness, and the ability to "see" and respond intelligently to the environment, exists in the simplest cell and probably has a lot to do with how it builds its metabolism and hence the origin of life.

The basic idea here is that DNA creates the cellular metabolism, and the proteins to facilitate it, in response to what the Halo "sees." This explains why cells so often are adapted in such a successful way to their environment. There are in fact several examples of cells in biological organisms "seeing" and directly adapting to changes in the environment. Consider, for instance, the differentiation of cells in a multicellular organism such as a human being. The stem cells develop into their particular cell types, not because of random mutations but because they are able to "see" their environment (which in this case is their neighboring cells) and adapt to this. Consider also the immune system, where the cell has an ability to "see" foreign substances in the bloodstream and tell the DNA what kind of antibodies it should generate to protect itself. We could go on at length to describe other examples of how the expression of the DNA is determined by what membrane proteins see. The point I am making is that this "seeing" by the cell and the response by the DNA are not merely the interactions of inanimate molecules, but an expression of the consciousness that any cell has because it was created in the image of the Cosmic Tree of Life.

While it is easy enough to see that single cells living today are conscious and are able to "see" their environment, it is more difficult to provide any direct evidence that cellular metabolism in the early formation period of cells, 3.8 billion years ago, was the result of the consciousness and a corresponding level of "intelligence" of the DNA. Needless to say, such a view is not consistent with that of the scientific mainstream, which essentially sees evolution as a product merely of physico-chemical reactions in which consciousness plays no part.

Yet in the case of bacteria, which are most relevant to the discussion of early life on Earth, an interesting case in point was reported by Cairns and coworkers in *Nature* in 1988.[4] These researchers studied mutations in an *E. coli* bacterium that had been deprived of lactose, a necessary nutrient, and they found that under these conditions this cell "produced" a mutation that compensated for the absence of lactose. Especially dramatic perhaps was the finding that no unselected mutants appeared in the experiments, only those that formed in response to the selective agent. This in the language used here attests to the ability of the bacterium to "see" the nutrient and respond immediately. This paper resulted in quite a controversy, since it introduced the concept of "directed mutations," which put in question the Darwinist dogma that biological evolution has no direction.

In further studies it has been found that indeed certain kinds of stressful situations (in which there is some external factor that the cells "see") may put organisms in what is called a hypermutable state.[5] In such a state the chance, or risk, that their genetic makeup will undergo change increases, which may make the cells more adaptable to the situation through mutation. The existence of these directed, or stress-induced, mutations as well as the altered mutability of bacteria are quite consistent with the proposal made here that cellular Halos may "see" and respond directly and "intelligently" to local physico-chemical conditions. This may also have been true in the stressful environment of early Earth, when the metabolism of cells first came into existence.

The kind of bacterial responses described here, which are not the result of mutations, are really expressions of what may be called Neo-Lamarckism, since the hypermutability emerged as a direct response to a need for change in the organism, and not as a selection of randomly generated variants, as Neo-Darwinism postulates. While Darwinian mechanisms for adaptation in microbes do exist, which is known from the development of resistance to antibiotics, the Neo-Lamarckian mechanisms, relying on the consciousness and intelligence of cells, presumably played a much greater role in creating the original biochemical processes of the first cells.

I want to emphasize here however that Neo-Lamarckian, or epigenetic, mechanisms refer only to the adaptations of cells to their environment and not to their evolution. This is important to point out because epigenetics has become popular in the recent decade as researchers have grappled with the inconsistencies of Darwinism. *Yet neither an epigenetic nor a genetic mechanism can alter the wave forms emanating from the Cosmic Tree of Life, from which the organisms are created. Epigenetic and genetic mechanisms serve only to adapt an organism to the environment within the constraints of its wave form.*

These are important distinctions to make in order to grasp how the theories of Neo-Lamarckism and Neo-Darwinism on the one hand differ from the Tree of Life theory on the other. Both of the former mechanisms, but epigenetics probably more so, may describe how cells adapted to their environment as life first originated, and also later in evolution. Yet, and this is very important, neither theory is capable of explaining the overall direction of biological evolution or what drives this. Jean-Baptiste Lamarck's idea that biological organisms develop their traits in response to how much they use them, and because of some internal striving, holds only a very limited truth. Changes, genetic or epigenetic, in a species can take place only within the limits constrained by the particular wave form that has defined the species and from which it was created. Thus, in general terms, even if organisms are conscious they do not evolve in the direction that they themselves might be striving for, but in the direction defined by the Cosmic Tree of Life. By themselves, neither Neo-Darwinism nor Neo-Lamarckism—neither genetic nor epigenetic mechanisms—can explain the evolution of cells. Both of these theories from the early nineteenth century suffer from the significant shortcoming that they look upon evolution as something that emanates from the biological organism itself. Thus neither theory has the necessary cosmic context for understanding evolution, which was unknown in scientific terms in the early nineteenth century.

THE PANSPERMIA OF CELLULAR WAVE FORMS

The traditional, and so far most commonly embraced, variant of the panspermia ("seeds everywhere") hypothesis proposes that life on Earth did not originate here, but was seeded from space in the form of spores or microbes. This idea goes back at least a hundred years to the Swedish scientist Svante Arrhenius[6] and has been advocated by many scientific notables, including Fred Hoyle and Francis Crick. A somewhat wilder version proposes that life on Earth was seeded by extraterrestrials in spaceships from civilizations more advanced than our own.[7] The panspermia theory may to some extent be seen as a response to the failure of Darwinism to explain the origin of life. Yet it is important to point out that, by itself, the traditional panspermia hypothesis does not explain the emergence of spores or microbes either here on Earth or elsewhere. Nor is it clear why life would be more likely to emerge somewhere else than here, especially considering that we have found that our Earth was tagged for the emergence of life.

Nonetheless, given that we have found that life on Earth has a cosmic origin, it is easy to understand that the theory of panspermia seem intuitively attractive. The report in 1996 of a meteorite from Mars with an age similar to that of the solar system, apparently carrying nano-bacteria to Earth, also gave new wind to this theory.[8] As it is known today that extremophile bacteria may survive temperatures well outside the current range of normality on Earth,[9] and that meteorites may travel between planets, the idea that seeds could have traveled to us through space does not in principle seem impossible. Yet it should be clear that if the emergence of life depended on such meteor travel, its appearance anywhere would be extremely precarious and it would certainly not coincide as closely as it does with the significant quantum shifts described by the Mayan calendar.

I instead would like to propose as a more reasonable explanation for the origin of life on Earth a *panspermia of cellular wave forms emanating from the Cosmic Tree of Life*. Considering that we have now identified the origin of life in the wave forms of DNA, this seems much stronger than a panspermia of spores or microbes. A theory of a panspermia of cellular wave forms explains the origin of life not only here, but potentially also everywhere else in the cosmos. The panspermia of cellular wave forms, really wave forms of DNA, combined with the notion that they may see, and adapt to, different environments, could explain the origin of life anywhere in the universe. It may also serve to explain three significant enigmas. The first of these is how life could emerge so soon after the Earth became hospitable; the second is how microbes with an extraterrestrial origin seem to have been found; and the third is how microorganisms could have emerged that survive under utterly extreme physico-chemical conditions.

To highlight the first of these enigmas, it has been very difficult to understand why cellular life appeared on Earth at such an early point in its history, almost as soon as the temperature became low enough to allow for this.[10] The oldest rocks on Earth found so far are the Acasta Gneisses in northwestern Canada near the Great Slave Lake (4.03 Gyr) and the Isua Supracrustal rocks in West Greenland (3.7 to 3.8 Gyr); the oldest microfossils similar to cyanobacteria appear in the latter site.[11] Yet the meteor bombardment that repeatedly vaporized the early oceans of the Earth is believed to have made it a barren and inhospitable place until at least about 3.9 billion years ago.[12]

At this point in time, however, at the beginning of Day 6 in the Cellular Underworld, two processes were synchronized to generate the necessary physico-chemical conditions for the emergence of life. One was that the temperature of the Earth decreased so much that water would stay in a liquid form. The other was that the returning meteor delivered water and the lighter elements necessary for life. When this happened, the first microorganisms started to appear. DNA wave forms presumably had been emanating from the Cosmic Tree of Life for a long time, perhaps even from the big bang. This meant that at the beginning of Day 6, all that was required for life to materialize was that the physico-chemical conditions be right, and, as this happened through the two abovementioned processes, life emerged.

My suggestion in chapter 4 that primitive protocells existed at a very early point in the evolution of the universe may now start to makes sense. Thus when the Cosmic Tree of Life started to generate wave forms of DNA, the possibility for life to emerge must have existed everywhere. Where and when life would arise would then depend on to what extent the physico-chemical conditions were conducive for this.

These DNA wave forms with Halos were thus projected onto all the galaxies of the universe, including our own and our particular solar system. As the solar system solidified in the form of planets, these Halos would manifest cellular life in locations where life could thrive. Thus wherever the temperature and chemical composition in our solar system were conducive to cellular life, it would have emerged, whether on Mars or on the early Earth or elsewhere. Wherever there was fertile ground for such Halos, they would give rise to cellular life, sometimes in the form of extremophiles, even under very challenging physico-chemical conditions.

Such a panspermia of wave forms would be consistent with observations of microbes of a presumed extraterrestrial origin found on our own planet. Findings in the currently thriving field of exobiology indicate that life may have emerged also in other places in our solar system, not just on Earth. I'm referring to the somewhat controversial, yet mostly accepted, interpretations made of mineral structures and organic matter indicating the presence of microbes in meteorites. Special attention has been paid to the 4.5 billion-year-old meteorite found in Antarctica originating from Mars, ALH 84001, which NASA considers as evidence of microbial life on early Mars.[13] Another is the Tagish Lake meteorite, which landed in Canada in 2000[14] and which some suggest is older than the age estimated for our solar system. However, evidence of life associated with this meteorite seems more doubtful. Even so, if such observations on meteorites, indicative of life, prove to be correct, it would seem that cellular life emerged on Mars as soon as it solidified and that life may have been present elsewhere even earlier. This would be consistent with a panspermia of DNA wave forms existing from a very early point in the universe, even though for us clearly discernible cellular life can be verified only in the later phase of the Cellular Underworld. If this is true, unicellular life may, through the distribution of Halos, have emerged at a number of other planets and moons in our solar system with liquid water, where certain other physico-chemical conditions were also at hand.

Similarly, the Halo panspermia hypothesis may shed some light on the emergence of extremophiles. These are unicellular organisms that survive, and even thrive, under conditions that by normal biological standards seem utterly extreme. Hence there are cells that can reproduce at a temperature of 121°C,[15] that live in acidic environments with a pH of less than 0,[16] and that thrive in environments of high pressures in deep sea vents. Extremophiles developed special characteristics to deal with these extreme conditions. There are also several classes of organisms, such as deep-sea bacteria, that are not ultimately dependent on the Sun for their energy supply. It has been suggested that some of the earliest organisms of life were such extremophiles that emerged independently of photosynthetic bacteria.[17] The panspermia of DNA wave forms could explain not only that extraterrestrial life could exist, but also that different types of cells in a variety of extreme environments on the early Earth could emerge independently of one another.

If this is true, the panspermia of wave forms would imply that life originated independently in several different locations and different physico-chemical conditions on Earth and in the solar system. This would mean that Darwin's theory of a common descent of all species from one type of bacteria is not likely to be true. Because of the panspermia of wave forms, life is instead likely to have emerged independently in several places and times on our planet as well as in the solar system. That all cells have DNA and proteins in common does not prove that one descends from another. It only proves the universality and unity of life in a wider sense, which is that it always emerges in resonance with the corkscrewing wave forms emanating from the Cosmic Tree of Life.

We should not be surprised then that the attempts of Darwinists to draw their so-called tree of life, describing how all life has descended from one original bacteria, has not been very successful. Their proposed phylogenetic trees are constantly changing and are also directly challenged by the discoveries of extremophiles, which are apparently unrelated to other forms of bacteria. Such attempts have failed because the whole concept of common descent is flawed, and no phylogenetic tree of life actually exists. Cellular life emerged as a result of a panspermia of wave forms producing a variety of species all over the universe, including in a wide diversity of extreme environments on Earth. They all share DNA as a basic molecular component because they have a common origin in the Cosmic Tree of Life. We may again see how the very coming into our awareness of the Cosmic Tree of Life tends to dispel confusion and how the leaves continue to fall from the illusory phylogenetic tree of life created by Darwinism.

8

A New Theory of Biological Evolution

FROM UNICELLULAR
TO MULTICELLULAR ORGANISMS

In the previous chapter we have discussed what is usually referred to as the origin of life. It seems however that in a real sense there is no particular event that can be called the origin of life except for the emergence of the Cosmic Tree of Life at the big bang. In a broader sense, every new quantum state of the Cosmic Tree of Life can be said to represent a new origin of life, generating a new range of species with a new level of consciousness. A very fundamental new quantum state occurred at the beginning of the Mammalian Underworld, which prompted the origin of the multicellular organisms. This quantum shift meant that the Cosmic Tree of Life began to project a completely new wave form, one that introduced a senior level of control of evolution, to which the individual cells in the multicellular organisms had to adapt.

The step from the Cellular to the Mammalian Underworld, a shift that led from the unicellular to the multicellular organisms, is one of the most significant in the history of biological organisms. This step also meant, as we shall see, that DNA was dethroned as the chief manifestations of the wave forms driving evolution. While DNA generates the proteins that facilitate cellular metabolism in individual cells, we are aware of no mechanism that could create the knitting pattern between the cells in a multicellular organism. We must ask new questions regarding the origin of such organisms. How may such a knitting pattern be created in a multicellular organism? In the established paradigm the standard answer is that the knitting pattern is coded by the DNA molecule, and this is taken so much for granted that it is rarely even discussed. Biology textbooks and the media alike typically say that DNA is not only "the code of life," but also the "blueprint of our bodies." In reality, all we know is that DNA is a sort of library for how cellular proteins are made, and this does not in any sense mean that it describes the anatomy of an animal. Before I suggest an alternative answer to how anatomies are created, I thus feel it is important to show several lines of evidence that DNA indeed does *not* determine the body plans of multicellular organisms. Because the notion that new multicellular species emerge through randomly induced changes in DNA is part of the Neo-Darwinist model, it is all the more important to show that this cannot be the case. These lines of evidence include: (1) There is no proportionality between the complexity of an organism and the size of its DNA or the number of its genes. (2) The degree of similarity between the actual DNA sequences in human beings and other species does not seem to reflect the degree of similarity in the anatomy and morphology of these organisms. (3) The information content in DNA is insufficient to define the cellular knitting pattern of a multicellular organism.

1. As the structure of DNA and the genetic code were elucidated in the fifties, researchers started to measure the amount of DNA that was present per cell (called the C-value) in different organisms. To their surprise they found that this C-value did not correlate particularly well with the complexity of the organisms in question.[1] If DNA provided the code for the body plan of a species, it was of course expected that

the DNA would be considerably larger in organisms with a complex morphology than in those with a simple one, and the absence of such a correlation came be called "the C-value enigma."

Today, more refined techniques have been developed to allow researchers to assess not only the total amount of DNA, but also the total number of genes (the so-called genome) in different organisms. As these techniques began to be applied, it was again expected that the more complex organisms would have the highest number of genes. The sequencing of the genomes allowed for the estimation of the number of genes in organisms such as the fruit fly (13,379 genes), the dog (19,300), a nematode (*C. elegans;* 19,427), a sea urchin (23,300), a puffer fish (27,418), and rice (37,544).[2] Consider these numbers in light of the billions of dollars spent by the Human Genome Project to determine that the human genome contains 20,488 genes,[3] which is fewer than in a sea urchin and commensurate with a nematode of less than a millimeter (see figure 8.1). If the genome size reflected the complexity of the body plan, we would expect the human genome to be several orders of magnitude larger than that of a simple nematode or even a dog. This is obviously not the case. There is thus nothing in the actual data that indicates that the multicellular complexity of an organism or its anatomy is reflected in its genes, and one would be hard-pressed to consider the morphology of a nematode as equally complex as that of a human being.

Given these findings by the Human Genome Project, it would have been natural to throw out the window the idea that DNA determines the morphology of multicellular organisms, or at least to start considering alternative explanations. This however did not happen, maybe because for so long it has been presented as fact that DNA determines our traits and morphology. Yet it is difficult to see how the fundamental view that DNA determines all of our biology, taught as fact to schoolchildren all over the Western world, can be reconciled with the finding that a nematode has an equal number of genes as a human being (the nematode in question has 969 cells and a human being an estimated 40 trillion).

Homo sapiens

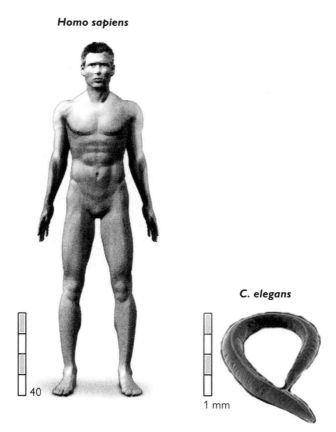

C. elegans

40

1 mm

Figure 8.1. Complexity difference between a human being and the nematode C. elegans. *According to the current paradigm of biology, DNA determines the anatomical and morphological complexity of an organism. If this were true, we would expect more complex organisms, like humans, to have a higher number of genes than simpler ones. In reality, the number of genes in the simple nematode* C. elegans *is about the same as in* Homo sapiens—*in direct conflict with the ruling paradigm.*

2. Even more refined methods have been used to directly compare the genomes of different species in order to study what proportions of their genomes are identical (or very similar). These studies have shown that human beings share 99.4 percent of their DNA sequence with chimpanzees,[4] 90 percent with rats, 89 percent with mice, and 83 percent with zebra fish. Again the results clearly challenge the notion that information

in the DNA determines the complexity of the body plan of an organism. If you believe that the degree of similarity of DNA sequences is indicative of the overall similarity in morphology, these results are not at all what you would have expected. Hence, humans and chimpanzees are certainly not look-alikes even though their genomes are practically identical, and most everyone would agree that there is more than a 17 percent difference in the anatomy of a zebra fish and that of a human being.

Naturally, work is currently being carried out to identify the actual DNA sequences that differ between humans and chimpanzees,[5] and the hope is to identify some genes, and their corresponding proteins, that could conceivably explain the great difference in morphology and intelligence between the two primate species. Yet my own sense is that an unbiased inspection of any difference that may be discovered in their respective DNA codes will turn out to be irrelevant when it comes to explaining the great differences between our own and other species in morphology and abilities.

3. The conclusion that DNA does not code for the anatomy and morphology of a multicellular organism is also supported by the fact that the information content in DNA is not sufficient to define the knitting pattern of an organism's cells. If we use the human being as an example, its total genome (including noncoding segments) holds approximately 3.5×10^9 base pairs of DNA in a linear order, while the adult human body holds trillions (40×10^{12}) of cells arranged in a specific three-dimensional structure. Could a DNA sequence of this size determine a knitting pattern of this complexity? To assess this we may for simplicity ignore the three-dimensional aspect of the knitting pattern and compare the linear DNA sequence with a simple hypothetical linear arrangement of all the cells of the human body. Since there are 4 different bases in a DNA strand and 260 different cell types in the human body, the number of possible DNA sequences would be $4^{3.5 \times 10^9}$, while the number of possible linear arrangements of the cells is $260^{40 \times 10^{12}}$. These numbers reflect an astronomical difference between the number of possible linear patterns of the cells, which incidentally would have

been even greater if three-dimensional coordinates for the cells had been considered, and the number of possible linear arrangements of the DNA bases. The conclusion to draw from this is that the information content in the DNA is totally inadequate for determining the anatomy and morphology of a human being (and presumably most other multicellular organisms as well).

Maybe it will come as a surprise to many that the existing evidence from the Human Genome Project is in direct conflict with the idea that the body plan of multicellular organsisms is coded for by DNA and that changes in this molecule would be driving its evolution. After all, since you inherit certain traits of physical likeness, such as the color of your eyes, from the DNA of your parents, it would seem only natural that the whole body plan be defined by this DNA. And yet, as we have seen above, this obviously is not the case. To understand why it is not, we need to realize that there is a difference between the actual body plan of an animal, such as the anatomy of a human, and factors that may modify this body plan, such as certain proteins coded by the DNA. Hence while DNA clearly may modify the expression of certain physical, and even mental, traits in humans, and such modifications may be heritable, it does not determine whether these traits exist to begin with.

Why then have so many biologists implied that DNA carries the entire blueprint for life? Apart from the ideological reasons it is also that there are certain mutations in DNA that seriously perturb the development of the body plans of multicellular organisms.[6] Hence if you irradiate insects to produce mutations, you will find that their legs sometimes grow out at the wrong places or that their wings may be malformed. Yet this only shows that some cellular protein that was necessary for maintaining the morphology of the organism was modified in such a way that malformations emerged. It does not prove that the particular gene "codes for" that morphology.* We may make the simple analogy with

---

*It is common for media reports to talk about different genes that "code for" certain traits and diseases. This wording is really misleading, since all that these genes do is specify the instructions for how a protein is to be made.

a car that by mistake has been endowed with an electronic part that is defective (corresponding to a mutated protein). While this could easily result in a whole system in the car being defective, this does not mean that the blueprint for the car was defective. It means only that for the proper functioning of the car, no parts can be defective.

Based on this evidence, it is possible to make a strong case that changes in the genes and their DNA do not play a significant role in the development or evolution of the overall body plan of multicellular species. DNA plays a crucial role as a library of blueprints for cellular proteins, and during the course of evolution this may have been modified to adapt to new environments, including that of neighboring cells in multicellular organisms. However, the DNA codes do not define the knitting patterns, anatomies, or morphologies of multicellular organisms. This naturally raises the question as to what actually defines the body plan. To address this we need to begin with the conclusion drawn in chapter 4 that animals, and human beings in particular, have been created in the image of the Cosmic Tree of Life. We must ask how it is that the morphologies and anatomies of animals have evolved to reflect the Cosmic Tree of Life, including their bilateral and erect body plans. If we can understand the mechanisms by which this has taken place, we would be able to explain how the body plans have evolved.

THE ORGANISMIC WAVE FORM

In chapter 4 it was shown that animals in general, and human beings in particular, are created in the image of the Cosmic Tree of Life. It was also shown that multicellular organisms in the Mammalian Underworld evolved Day by Day into symmetrical organisms that increasingly reflected the Cosmic Tree of Life, with lateralized eyes and brains for processing sensory information. It was then also pointed out that such a directed evolution could not possibly have been brought about by random mutations in the DNA. We have yet to discuss, however, the mechanism by which the evolution of multicellular organisms *from the very onset* of the Mammalian Underworld, 820 million years ago, has been directed toward creating mirror images of the Cosmic Tree of Life. To understand this we will again need to consider how the Halos on different levels—galactic, planetary, and organismic—serve as a series of lenses, as we saw in chapter 6. In the model proposed here, these lenses project a wave form down to the level of the multicellular organisms that are created by the Mammalian Underworld. As we may also recall, as the Mammalian Underworld was activated, a 90-degree shift in its dominating yin/yang polarity took place in accordance with the Cosmic Round of Light. This resulted in the emergence of wave forms that were primarily left-right polarized about their midlines. It is this polarized field that is being projected down through the series of lenses to the wave forms at the multicellular level (referred to as the "organismic" level), and it is because of this mechanism that the Days of the Mammalian Underworld tend to give rise to animal species whose higher forms are bisected and retain a lateralized form. Through the repeated introduction of new yin/yang polarities in the Mammalian Underworld the wave forms were driven to increasingly generate biological organisms that mirrored the Cosmic Tree of Life.

An important rule in this model of the evolution of the universe is that the higher levels of organization of life are senior to the lower, and that the latter always need to function within the frameworks provided by the higher levels. This rule also applies to the relationship between a multicellular organism, such as an animal, a plant, or a fungus, and its individual cells. Thus in a healthy organism, the functioning of individual cells is subordinated to the body organs they are part of, which in turn are subordinated to the whole organism. There is in fact a name for what happens when single cells do not follow this seniority rule, and this is cancer. In the established paradigm of biology, there is currently very little understanding of what ultimately structures the whole organism and makes the individual cells adapt to the whole, but in the current model this will be explained by the application of the seniority rule to entangled systems on different levels.

It is also important to point out that the Halos on our own particular organismic level have certain characteristics that are unique. First, while previously we have seen how planets came out of star systems that came out of galaxies that came out of the cosmos, the organismic level is actually created from a *lower* level of organization of life, the cellular. In fact, the organismic level may be looked upon as an interface where the projection of wave forms from higher levels of the cosmos orchestrates the creation of multicellular organisms, starting from single cells. Second, because of the creation of animals in the image of the Cosmic Tree of Life (which, as we have seen, the entire evolution of the Mammalian Underworld is all about), the organismic Halos are not, unlike at all other levels, spherical in nature, but more like cone-shaped auras. Third, our own Tree of Life, essentially the spinal cord and its extension in the brain, is not perfectly straight and more notably does not rotate, something that apparently distinguishes it from Trees of Life at all other levels.

That there are certain peculiarities associated with our own level of existence, and the wave forms that have shaped it, should however be no cause for surprise. In the current theory we are not out to prove that the human being is an immaterial whim of nature. On the contrary, we have shown that the human being is the intended end result of a purposeful evolution and have reason to take an interest in the characteristics that make us special. Hence I think it is a basic premise that the purpose of this creation is not only to create an organism in the image of the cosmos, but also to create one that is conscious and is able to experience reality. I feel it is because organisms on our own level are meant to be conscious that we do not rotate and that we, unlike the plants and fungi, also have the capability to move around freely, physically speaking. Our biological structure, and the organismic wave form that shaped it, is thus the result of a directed evolution aiming to create a being that can experience the richness of reality.

For a directed evolution to take place, there must exist a template for the morphology and anatomy of an organism, and as argued above this template is the Cosmic Tree of Life, projected down through the lenses of the Halos of the Cellular, Mammalian, Anthropoid, and Human Underworlds. The resulting wave form provides a template for the *adult* organism and is what drives embryonic and fetal development. (In biology, the term *development* refers to the changes an individual member of a species undergoes in its lifetime, whereas the term *evolution* is reserved for the long-term changes of different species.) Ultimately the design of a biological species, and the design of its overall morphology and physiology, is generated by this organismic wave form. This wave form is also behind the specialized biochemical mechanisms of its individual cells, which are subordinated to the functioning of the whole organism. It is this design, provided by the organismic wave form, that in a specialized cell determines what is expressed in the genome. This model for cellular differentiation and morphogenesis in the development of an organism explains the observation made a long time ago that ontogenesis (the development of a single member of a species) reflects phylogenesis (the evolutionary history of a species). Thus when a fertilized cell is activated by an adult organismic wave form, it initiates the development of an individual organism, which in some ways reflects the evolution of the species brought by the same wave forms, maybe throughout the course of several Underworlds.

However, in order for the organismic wave form to manifest itself in the form of a higher animal, a number of conditions must be met with regard to the individual cells. The first is that they must be eukaryotic cells with a diversity of organelles that may provide for an inner differentiation of cells. The second is the presence of a cytoskeleton, primarily constituted by microtubules, which among other things have the potential to generate different cell types by altering the shapes of the cells and in the positioning of their organelles. Third, at the center of the centrosome that organizes the cytoskeleton there must be an organelle

called a centriole. Fourth, the inner core of the Earth must have solidified in such a way that its Tree of Life may contribute to the formation
of a morphogenetic field around the Earth. If one of these conditions is
not met, higher bilateral organisms could likely not evolve.

These conditions seem essentially to have been met on Earth at the
beginning of the seventh Day of the Cellular Underworld, when the
eukaryotes emerged. The prokaryotic cells that had emerged earlier as
manifestations of the cellular wave form of the sixth Day of the Cellular
Underworld, however, did not meet any of the above conditions and so
did not have the potential to evolve into complex multicellular organisms. This is an example of how the seventh Day of an Underworld
often provides a platform for the development of a new Underworld on
top of it. In this case, the eukaryotic cells provided the necessary foundation for the evolution of the higher organisms in the Mammalian
Underworld.

THE CYTOSKELETON AND THE DEVELOPMENT OF MULTICELLULAR ORGANISMS

The step from a unicellular to a multicellular organism is quite large,
and we need to look at how the prokaryotic cells of the sixth Day were
transformed to meet the above conditions. Most notable among these
changes was the inner compartmentalization of these cells to include
different organelles, such as a cell nucleus with chromosomes, which
resulted in a eukaryotic cell. As mentioned earlier, the increase in
complexity of multicellular organisms requires that these be made up
of several different cell types, and what accounts for the variation of
these different cell types is their varying content of organelles and how
these are positioned and relate to one another within the cells. The differentiation into different cell types with specialized functions allows
for a meaningful distribution of labor among the individual cells so
that organs with specialized functions, constituted by such specialized
cells, may be created as part of a multicellular organism. Differentiation

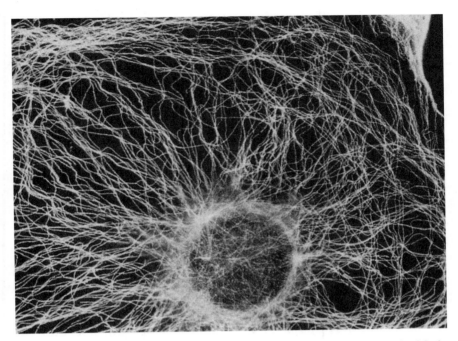

Figure 8.2. The microtubular cytoskeleton. Fluorescent antibodies highlight the microtubules (and nothing else) in this picture, which shows how the cytoskeleton in a mouse cell radiates from the circular centrosomal region. With permission from Mary Osborn and Klaus Weber, Cell 12 *(1977): 563.*

into different cell types also required the emergence of a new type of cytoskeleton in the eukaryotic cells, which was made of microtubules* (figure 8.2).

To understand how biological organisms evolve on a larger scale, it is then relevant to first briefly study how an individual organism develops from a fertilized egg to its birth. After an egg cell has been fertilized by a

---

*Until about ten years ago it was generally considered as fact that prokaryotic cells did not have a cytoskeleton. In recent years this has changed, and homologs have been found to most of the components of the eukaryotic cytoskeleton (see Yu-Ling Shih and Lawrence Rothfield, "The Bacterial Cytoskeleton," *Microbiol Mol Biol Rev* 70 [2006]: 729–54, doi:9.1128/MMBR.00017-06). Less is known about the structure and components of this proteotic cytoskeleton, but it is clear that the eukaryotic cytoskeleton is made up of components other than the bacterial.

sperm cell, it undergoes a series of rapid cleavages that lead to the formation of a blastula. This sphere is not larger than the original egg cell. As
the blastula is transformed into a gastrula, different layers of cells in the
embryo called ectoderm, mesoderm, and endoderm begin to differentiate into different cell types. From these different layers the shape of the
fetus emerges through a series of convolutions that develop its morphology through the growth, and continued differentiation, of cell types until
birth and through to adulthood. As part of the developmental process
of an animal, it should be noted that cells may also move and migrate
to different positions, where they then may be committed to specialized
functions. In this regard, the developmental process of animals is different from that of plants. On the Internet and elsewhere animations are
available that show human development from a fertilized egg to a fetus,
which may be useful for studying this process. At about twelve weeks of
pregnancy, the basic morphology of a human being with all its different
organs is developed, even though the fetus continues to grow rapidly.

The established paradigm in biology has it that this total transformation from a single fertilized egg to a mature individual is defined by
the genetic program of the DNA. This is then supposedly expressed
in the form of morphogenetic determiners, presumably proteins such
as growth factors and membrane receptors, that transform the linear
data of the DNA into information that describes three dimensions of
space and time and so generates the morphology and anatomy of an
organism. In a previous section we discovered however that all the evidence speaks against the genome playing such a role in determining the
morphology and anatomy of a multicellular organism and that its information content would be insufficient to define a body plan. Moreover,
as we studied the evolution of animals in chapter 4, we saw that the
evolution of the anatomies that emerged in the different Underworlds
was fundamentally related to a three-dimensional coordinate system. It
seems clear that the basic directions of a higher organism and its axes
of polarization—left-right, anterior-posterior, and dorsal-ventral—are
already established at an early point in embryonic development. No

explanation has however been provided as to what may be the origin of these axes in the early embryo and so provide the outline of a structure for the further development of an organism. In fact, as is so often the case, the relevant questions are not even asked, because it is taken for granted that the answer is in the DNA.

Given their roles in evolution, it is necessary to understand what provides the alignments for these basic morphological axes. What is at the origin of these is far from obvious, as in the physical universe, except for in some crystals, straight lines exist only as abstractions constructed from the spins of the polar axes of the cosmos, galaxies, stars, and planets, that is to say the Trees of Life on these levels. Given that we saw in chapter 4 that every Day of the Mammalian Underworld meant a step toward a more marked left-right polarity, and in fact three-dimensionality, of the animals that it gives rise to, and that this was related to the Tree of Life on different levels, we will here assume that the alignments around which a fetus develops originate from the Cosmic Tree of Life.

If the axes for the development of an organism in three dimensions are provided by its Tree of Life, we need to ask how an individual fertilized egg comes to be aligned with these dimensions. To see this we must look a little bit more closely at the cell and return to the fact pointed out above that for the development of the animal fetus it is necessary that some of its cells change shape and migrate to different positions. Individual eukaryotic cells move essentially like amoebas that extend pseudopods in the direction of movement, or in other words, by changing their shape. In embryonic cells this movement is orchestrated by the *cytoskeleton,* which is the "skeleton" of the cell (figure 8.2). The name cytoskeleton is however not very appropriate, since we tend to associate it with something that provides a firm structure to our bodies. When it comes to the skeleton of the cells, the cytoskeleton, its functions are at times the very opposite of this, as it is a very dynamic and flexible structure. It may undergo much change, and these changes are what transform the architecture of the cell. The cytoskeleton may also sometimes be likened to a structure of ropes

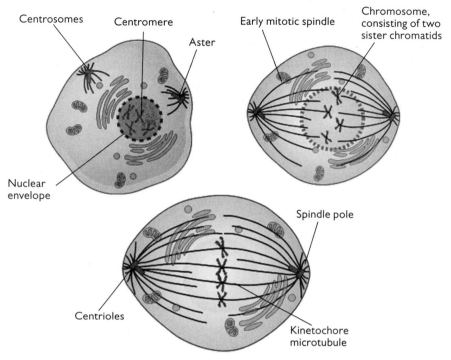

Centrosomes   Centromere   Early mitotic spindle   Chromosome, consisting of two sister chromatids

Aster

Nuclear envelope

Spindle pole

Centrioles

Kinetochore microtubule

Figure 8.3. The role of the centrioles in cell division. The replication, movement, and positioning of the centrioles play the decisive role in organizing and directing cell division. The microtubules emanating from the centrioles are pulling the chromosomes in opposite directions as part of the process of creating two new cells. The centrioles play a directing role at each stage in the cell division process.

that attaches to the organelles and pulls them in different directions, and this we may see in cell division when the chromosomes are pulled apart (see figure 8.3). The major components of this cytoskeleton are called microtubules, which are long tubes formed by thirteen filaments made from two alternating proteins known as alpha- and beta-tubulin (figure 8.4). Among the many roles of these microtubules are transporting and positioning different organelles appropriately for the functions associated with a special cell type. It is also to form and maintain the structure of the respective cell functions. Hence the cytoskeleton plays a crucial role in the differentiation of cells by commiting them to different functions.

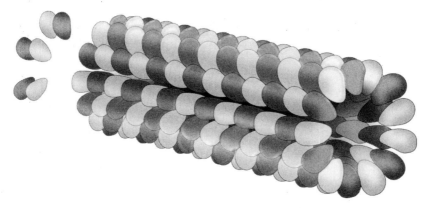

Figure 8.4. The structure of microtubules. Microtubules are made up of thirteen filaments of two alternating proteins, alpha- and beta-tubulin. They may grow at one end by adding combinations of the two proteins, called dimers, and shrink in the other end as dimers go back into solution. Courtesy of Bengt Sundin.

Figure 8.2 shows how the microtubules that direct many of these processes emanate from a microtubule-organizing center. The cytoskeleton also plays a crucial role in the movement of individual cells and may serve to transmit signals between them.[7]

Located in the center of the microtubule-organizing center is an organelle called the centrosome. The centrosome was discovered in 1888 by the great German biologist Theodor Boveri,[8] and it was given its name because of the central organizing position that it seemed to occupy in the cell. In fact, until Neo-Darwinism shifted the focus of attention over to DNA in the 1950s, biologists typically looked upon the centrosome as the regulatory center of cells, and this is still a natural conclusion to draw if one has no preconceived ideas. To illustrate such a regulatory role, we may see that there are two centrosomes that initiate cell division by attaching microtubules to the chromosomes and then pulling them apart in opposite directions by a mechanism in which the microtubules are shortened (figure 8.4). This means that the direction of growth of a tissue, and *the relative positions of* its *cells, is determined by the centrosomes,* since those determine the direction of cell division.

As mentioned above, the cytoskeleton emanating from the centrosome also commits the cells to different types through the internal positioning of their organelles.

Moreover, and equally important from a developmental point of view, the centrosome directs the external movement of a cell by attaching its microtubules to the walls of the cell in crucial positions and then expanding and contracting the microtubules. It is actually the microtubules of the cytoskeleton that through such attachments give the cells their particular shapes. In short, *the shape, type, and movement of an individual cell are determined by the microtubules of the cytoskeleton emanating from the centrosome.* It seems clear then that the centrosome and the microtubules of the cytoskeleton play roles of paramount importance in the life of the cell and in fact determine its morphology. Since an organ is created from (a very high number of) such cells, its morphology reflects the sum of the morphologies of all of its constituent cells. These then ultimately hinge on how the centrosome regulates the cytoskeleton in individual cells of the organ. The combined mechanisms of the centrosomes in all of the cells actually determine the crucial factors for the development of an animal from the fertilized egg to the mature adult organism.

Since this has been a discussion in general terms of the relationship between the cytoskeleton and development, it may be useful to look at an example of the actual role of the microtubules in the development of an organism. A good example is the development of the notochord in the early embryo (figure 8.5). The notochord is an embryonal feature shared by all higher animals, including ourselves, and it is the earliest manifestation of the spinal cord and brain, or, in other words, the central nervous system. Hence it is a relevant feature for the purpose of our discussion, since the axis of the spinal cord is the chief axis around which the three-dimensional morphologies of all bilateral organisms are developed. The notochord provides an early form of the Tree of Life as this is manifested in the development of higher animals. It turns out that this embryonic form is created by the shifting of the shapes, and convolutions, of a number of cells in the ectoderm, whose microtubules

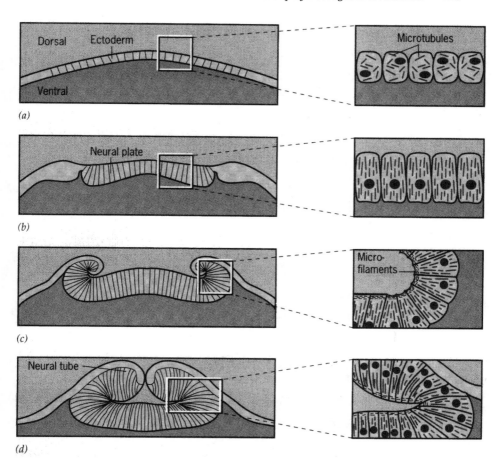

Figure 8.5. Stepwise formation of the notochord (later to become the spinal chord) in a chick embryo. In the ectodermal layer (a) of the embryo the microtubules emanating from the centrosomes align themselves with the direction determined for the notochord (b). The individual cells then migrate and change shape in relation to this alignment (c and d) so that an organ is formed, which provides a backbone for the three-dimensional organization of the body. Used with permission from Gerald Karp, Cell and Molecular Biology, *2nd ed. (New York: John Wiley and Sons, 1999).*

align with the wave-form axis for the notochord-to-be. Since the microtubules emanate from the centrosomes, it appears that the microtubules in all of these cells were affected by some kind of change in the morphogenetic field. This then initiated the changes in the shapes of these cells and their movements, and so gave rise to this crucial embryonal organ, the notochord, around which our bodies are later lateralized. From this example it seems that the ultimate reason that we have a central nervous system around which our bodies are organized in three dimensions, front-back, left-right, and up-down, is that the microtubules of the cytoskeletons of the participating cells at an early stage of embryonal development through Halographic resonance were directed in a special way by their centrosomes. It is in fact only logical that on a cellular level all major morphogenetic events are based on crucial alignments emanating from the centrosomes of the affected cells, alignments that at later points in development are related to the original notochord formation and the creation of bisected polarities around it. *All development in multicellular organisms, including their anatomy and morphology, ultimately emanates from the centrosomes.*

THE CENTRIOLE AND ITS ROLE IN EVOLUTION

Biology has long been haunted by the possibility that the primary significance of centrioles has escaped them.

DR. PITELKA (1969)

Of what is this centrosome, which seems to be at the center of all development of higher biological organisms, composed? It is primarily made up of a somewhat fuzzy pericentriolar material surrounding a much more structured organelle, which is called the centriole (see figure 8.6). We will here primarily focus on the centriole, whose structure with remarkably little change seems to have been preserved throughout almost a billion years of evolution of multicellular organisms.[9] For a relatively long time this spectacular organelle, which in its cellular context can be seen in figure 4.1, has not however received much scientific attention. Although some interesting studies have recently been made of the centriole, no monograph exclusively focusing on it has been published since 1982.[10] This illustrates the old truth that we see only what fits into the dominating worldview, while things essentially are ignored if they do not. While the location of the centriole in the midst of the centrosome has been known by researchers for more than a century, and the details of its architecture since the advent of electron microscopy, some have concluded that the centriole does not have a vital function.[11] This has been argued because the centriole is not indispensable for cell division in any species of plants or fungi, or in cell lines cultivated in the test tube. Considering that the centriole has existed in animals for almost a billion years, we have reason to believe that many biologists are not looking in the right direction for understanding the role of this organelle.

The very structure of the centriole immediately strikes the eye as fascinating and looks like the result of an intelligent and purposeful design. It is composed of two parts, each of which is not only straight, but perpendicular to the other as well. The centriole is predominantly

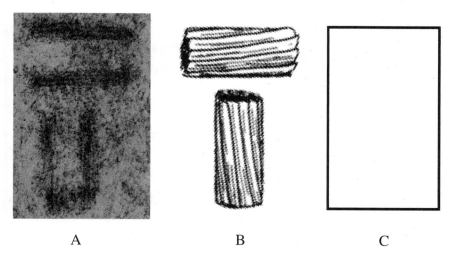

A   B   C

Figure 8.6. The centriole. (a) A photograph of a centriole taken by means of electron microscopy from the cell of a fruit fly (Drosophila). Used with permission from Visuals Unlimited, Inc. (b) Stylized drawing of a centriole, courtesy of Stuart Hameroff. (c) A rectangle that provides a framework for the centriole.

made up of the same microtubules as the cytoskeleton, made mostly from alpha- and beta-tubulin proteins, but in contrast to the cytoskeleton, the centriole has a fixed geometric structure. By now, the reader has probably already figured out the fascinating direction in which we are heading and has wondered what the origin of this perpendicular structure at the center of the cytoskeleton may be. What is especially relevant is that the centriole seems to be the only structure in the cell that provides a straight line and a 90-degree angle that in principle defines a cellular coordinate system. The centriole may thus at the very least be considered a leading candidate for the role of chief organizer of the dorsal-ventral, left-right, and anterior-posterior three-dimensional coordinate system that all higher animals are built around. We may in fact ask ourselves how cells would be able to align themselves in special ways, such as we saw in the formation of the notochord, if there were no inner coordinate system to relate their alignment to. Hence without something like a centriole in its midst, a cell would have no internal coordinate system as a basis for the positioning of organelles internally or for defining its spatial relations to other cells. Without an inner coordinate system, it is in turn hard to see how symmetries and a basic three-dimensional structure could ever be developed by an organism. While it is true that organisms of several kingdoms do not seem to need this organelle, no one seems to have taken note of the fact that *the centriole is present only in bilateral organisms that are built around a fundamental three-dimensional structure with a left-right symmetry,* a symmetry that seems consistent with the very structure of a centriole. In fact, the very presence of an organelle with the perpendicular structure of the centriole would tend to introduce a polarity, at least on the cellular level, that when it is shared by many cells results in a polarity for the whole organism.

This and several other observations indicate that it is the centriole that is primarily responsible for establishing the three-dimensional structure of the knitting pattern of higher animals. Thus before each cell division, centrioles replicate in a process where a so-called mother centriole serves as a template for a daughter centriole, whose position in the new cell is also determined by the former. This replication has some very interesting characteristics, and some crucial findings regarding this have been reported in a recent paper by Feldman and coworkers. It states: "We show that the distal portion of the centriole is critical for positioning, and that the centriole positions the nucleus rather than vice versa. We obtain evidence that the daughter centriole . . . relies on the mother for positional information. Our results . . . suggest that centrioles can play a key function in propagation of cellular geometry from one generation to the next."[12] Because the mother centriole positions the daughter centriole, a network of geometric alignments between centrioles is created that through cell division may play a crucial role in determining the direction of growth (figure 8.3). Moreover, if the original centriole (or sometimes two of them) in the fertilized egg contributes the basic three-dimensional axes of the organism as a whole, this network may position and relate all the daughter centrioles to the original axes and hence establish the precise knitting pattern of the cells. This would amount to an excellent plan for ensuring the structural integrity of an organism in such a way that it is always aligned with its body plan and anatomy, and this is obviously exactly what is needed for the creation of a multicellular organism.

These findings tell us not only that the centriole, being the central component of the centrosome, plays a decisive role for the alignment of all cell morphology, differentiation, and migration, but that it also participates in the creation of a network of cells by coordinating their geometries. It has been shown that even in such a simple organism as the nematode *C. elegans,* the basic left-right polarity is established at the beginning of embryogenesis, and in mutants, where centrioles are absent, this dimensional axis does not appear in the further growth of the embryo.[13] Moreover, the proper positioning of centrioles has been shown to be necessary for establishing the left-right asymmetry in mammalian development, because they govern the orientation of cilia.[14] The data indicate that at the appearance of the first centriole in the fertilized egg, an overall three-dimensional coordinate system is established for the developing organism.

It should also be noted that the mother centrioles serve as templates for their daughters without the DNA having any intermediary role and that this molecule does not play a part in determining the geometric network of the developing organism. Earlier it was thought that centrioles might have emerged from a process of endosymbiosis, but this theory has now been abandoned.[15] Since centrioles do not have DNA of their own, it seems unlikely that they would have emerged in such a way. More profoundly, since they are known to position the cellular nucleus with its DNA rather than the other way around,[16] *the centrioles seem to hold the primary role in the organization of eukaryotic cells.* Because of this primacy of the centriole in relation to the DNA, I feel that it is logical to propose that it was the centrioles that brought about the endosymbiosis of all other organelles. This proposal has not been considered previously, but seems consistent with their location at the center of the cytoskeleton, which transports and positions the organelles.

In addition to replicating themselves and passing on their directionality to the daughter centrioles, the centrioles may also sometimes undergo a transformation to so-called basal bodies, which serve as templates for two other very significant organelles in eukaryotic cells, flagella and cilia. Examples of flagella are the tails whose movements drive sperm cells forward, while cilia are the little hairs on the outside of cells that help them move, or serve to move particles on their surfaces. An example of the latter is when coordinated movements of cilia on the surface of lung cells transport foreign particles out of the respiratory tract. While the centrioles seem to regulate the internal movement of the organelles, for instance during cell division, its derived organelles, flagella and cilia, bring about the external movements of the cells. In fact, any time individual cells display movements that appear intentional, or purposefully directed, we can be fairly certain that one of these organelles directly derived from the centriole is taking part.

Although the name *centriole* implies a central position in the cell, this organelle sometimes changes position not only when cells grow, but also when they migrate or participate in the healing of wounds.[17] They sometimes move to the surface of the cell to serve as basal bodies that are the anchoring points of cilia or flagella. The proper orientation of the centrioles is thus critical for the proper movement of cilia, which often move in long synchronized rows,[18] and organisms that are unable to develop such cilia will die.[19]

It should be pointed out that because of their perpendicular structure, the centrioles define a three-dimensional coordinate system for each and every single cell. Without such a system, and contacts between the centrioles of different cells, it seems unlikely that the almost perfect symmetry in the external anatomy of higher animals could have been created. How otherwise, in the absence of an internal coordinate system, would the cells "know" how they are directionally related to the large-scale anatomy of an organism and in what direction they should grow to manifest the development plan? Such a coordinate system would also seem to be necessary to direct the growth of the cells in an animal when tissues have been excised or damaged and need to be regenerated and healed.

To summarize, we have seen that (1) centrioles exist only in animals; (2) centrioles establish the basic left-right polarity in animals: (3) centrioles establish a network of individual cell geometries; and (4) centrioles may move to correctly position themselves, for instance to help the synchronized movements of cilia or to help in the healing of wounds in a way that appears "intelligent." Yet in the established paradigm the positioning of centrioles, which from our perspective seems like the most important factor determining the early development of an organism, remains a dilemma and their movements hard to explain. While a recent study suggests that centrioles seem to be moving in some kind of field,[20] the nature of this has not been elaborated, as the nature of this cannot be described within the ruling paradigm. In my own understanding, this field is the organismic wave form generated by the

Cosmic Tree of Life. What I am proposing, and what is consistent with all the data presented so far in this book, is that the positioning of all centrioles in a multicellular organism is subordinated to the body plan provided by the organismic wave form, which in turn is a projection from higher levels of the Tree of Life.

The centriole, which in the fertilized egg emerges from a basal body brought by the sperm cell, in the Tree of Life theory serves as the anchoring point for the organismic wave form, which defines a three-dimensional knitting pattern, which defines the ideal positional relationships between the cells of the body. The subsequent movements and positionings of the centrioles take place in order to integrate and align the cells in accordance with this knitting pattern. From conception and onward, the organismic wave form organizes the alignments of the centrioles in such a way that the morphology develops according to a pre-set plan to become the adult organism.

I am also proposing as part of the Tree of Life theory that the organismic wave form, through its effects on the centriole, may commit different cells to specialized functions. We have already seen that centrosomes, and the centrioles at their core, play a critical role in the differentiation into cell types because of their positioning and organizing of organelles. We have now arrived at a theory in which all aspects of the development of an organism in fact may be determined by its wave form and mediated by its centrioles. From the concept of an organismic wave form, which outlines the body plan of an adult organism, we may incidentally also understand why we remain the same person throughout our life. The integrity of our being is maintained by our organismic wave form.

For the development of individual organisms, all the pieces of the Tree of Life theory seem to have been accounted for, but the same relationships (that the organismic wave forms influence the centrioles, which informs the centrosomes to align the cytoskeleton to a certain morphology and position the organelles to commit to a specific cell type) can be used to explain the biological evolution of species as well.

In fact, I feel that for the first time we have now arrived at a theory that in a plausible way explains how sudden jumps—involving several synchronized transformations—may take place in biological evolution and result in the creation of new species. If the wave forms can bring about the immensely complex transformation of a fertilized egg to an adult organism through the mediation of the centrioles, why would they not be able to do so in the long term-process of biological evolution? Hence transformations of the cytoskeletons resulting in new cell types could conceivably take place very rapidly, and be synchronized with anatomical changes, as a new wave form emerges, since such transformations could be mediated by the centrioles. If this is true, we may understand why intermediate forms are lacking in the fossil record and how new species often emerge without any traceable ancestral forms. Changes in overall anatomy and morphology could in other words be synchronized with changes in cell types, which would make sudden transformations possible, something that would be impossible in the Darwinist model of slow, gradual, random change. What is more, if we can link the shifting alignments to the shifting yin/yang polarities of the Cosmic Tree of Life, we can also understand why the emergences of whole new classes of species with more developed nervous systems, greater intelligence, and left-right polarity indeed follow the periodicities of the Mayan calendar.

It may serve us better to use an example here, such as the transition of biological species from sea to land or the reverse transition of a land-based mammal to a whale or other cetacean. In both cases such transformations of species require not only a complete change in morphology and skeletal structure from fins to legs or from legs to fins, but also synchronized changes in the dermal, respiratory, and excretory systems, to name only a few. It has always been a weak point of Darwinism that it has not even in principle been able to explain how the synchronistic transformations of several different organ systems could take place, since it assumes that change in each organ system is slow and that every step gives a survival advantage to the whole organism. It cannot explain synchronistic changes in several different organ systems in a species, since most small changes lack survival value unless the whole organism is transformed. For the evolution of a land-based mammal to a whale, there is for instance no value in having fins if dermal and respiratory systems that are adapted to life in water do not emerge at the same time. Conversely, there is no selective value in having a skin adapted to life in water in the absence of fins for swimming. The different phenomena favoring life in water need to emerge in synchrony if the odds for survival are to increase, and the reason that the fossil record very often attests to sudden jumps where entirely new species emerge is that such a complete transformation of the organism is necessary for it to be able to survive the transition to a new environment.

The Tree of Life theory may however account for sudden jumps in the fossil record, since it can explain synchronistic changes taking place through the transformative effects that the introduction of a new wave form may have not only on the anatomy, but also on different organ systems and their cell types. As is typical of the Mayan calendar and its evolutionary progressions, synchronistic changes may be expected at any introduction of a new yin/yang polarity. Through this model it becomes understandable that a land-based animal may be transformed to a whale, even though this requires that several cell types and organ systems be synchronistically altered. The reason is that the yin/yang polarities are mediated by the centrioles on a cellular level. Changes in anatomy are synchronized by the cells, and in this way the necessary physiological transition, or sudden jump, becomes plausible. During evolutionary transformations, as well as during the development of single individuals, the whole system of metabolism has to adapt and be restructured to serve the new physiology and morphology of the larger whole. To effect this, the seniority rule determining the relationships between Halos at different levels of the hierarchy of life is necessary. The synthesis of new proteins by the DNA has to be subordinated to the overall transformation of the organism mediated through the centriole. Logically there can be no stepwise trial and error or gradual change in such evolutionary processes generating sudden jumps. What is proposed here instead is that a very dynamic process of transformation takes place where *the organism undergoes synchronized changes in its morphology, physiology, and biochemistry that could not have occurred in isolation from one another.* All of the parts are transformed simultaneously with the whole, and the proteins and genome are transformed to serve the whole, rendering DNA changes secondary. The development of a new individual is driven in much the same way, and from such synchronized changes we may understand why a fertilized cell grows to an embryo to a fetus to a child to a grown-up person. The reason is that it is the wave form of the *adult* organism that is merged with the centriole of a fertilized cell. The yin/yang polarity ensures that this cell undergoes the sequenced pattern of cell divisions, which is driven and molded by the adult wave form, to its preset "ideal," and the particular course development takes is subordinated to this overall purpose.

Since it allows for the quantum jumps often observed in the fossil record, it seems that in all respects the new theory of biological evolution has greater explanatory power than Darwinism. It explains not only when, but also why and how biological species evolve. Yet we must ask where this theory leaves the role of the genome. This question is especially important given the large body of work that demonstrates how the course of development of biological organisms can be substantially altered by mutations in certain genes. There are for instance thirteen Hox genes that, if mutated, result in substantially altered morphologies for the species in question. From stem cell research it is also known that embryonic cells from different species can be brought to differentiate to specific cell types by injections of various mixtures of growth factors that directly affect their genomes. Thus cultured stem cells seem to be able to differentiate without an organismic wave form.

Another fact that points to a significant role of DNA in development is that centrioles have been found to induce parthenogenesis (virgin birth) if they are implanted in frog eggs. To begin with, we should emphasize that this by itself provides significant and remarkable confirmation of the primary role this organelle has in development; for the development of a frog egg there is no need for sperm DNA, only for a centriole. Yet, what complicates the interpretation of this finding is that frog eggs develop into mature frogs regardless of whether the centriole is derived from a frog, a rat, or a human.[21] This tells us that there must be components in the frog egg that inform it that it should develop into a frog and not into a rat or a human, and these components, such as specific growth factors, most likely are coded for by the DNA. This shows that the knitting pattern of the cells and the body plan of an organism are not found "in" the centriole. Rather, the knitting pattern is mediated by the centriole, which directs the genome to be expressed in a coordinated way, but information in the DNA commits the development to a specific species. The genome may play an informative and modifying role in development, and this may be what we are seeing in these centriole-created virgin births. After all, we also see inheritance patterns of certain bodily, and often facial, traits that presumably manifest because both parents provide DNA to the fertilized egg, and this indicates that DNA may indeed modify the development of the body plan of an individual.

From this we may understand that even if DNA, and the various proteins that it codes for, plays a subordinated role for developing the three-dimensional anatomy of a multicellular organism, it nonetheless may be the source of crucial information. It also provides the cellular material and critical growth factors that the wave form interacts with. The point to realize however is that in a multicellular organism, cellular differentiation always occurs within the context of an overall plan for the development of an organism, and this is provided by the organismic wave form emanating from the Cosmic Tree of Life. This seniority of the organismic wave form is also the reason that the introduction of new yin/yang polarities, maybe especially clearly evident in the Mammalian Underworld, plays an overriding role in bringing biological evolution forward toward bilateral organisms that increasingly reflect the Cosmic Tree of Life.

9

A Biology with Soul

THE CELLULAR TREE OF LIFE

Nothing in Biology makes sense except in light of the Mayan calendar.

CARL JOHAN CALLEMAN (2008)

In several specific ways it seems that the centriole functions perfectly in accordance with what we would expect from a Cellular Tree of Life. It has a perpendicular structure that induces left-right asymmetries, which in chapter 5 we saw to be a characteristic of Trees of Life at higher levels of the universe. Also, this organelle cannot possibly have been formed by innumerable small, incremental, random changes in proteins that would then form tubulins and then microtubules and then centrioles, flagella, and cilia, which has previously been pointed out by advocates of intelligent design. What is even more telling regarding the centriole as a possible Cellular Tree of Life is its actual architecture. As I mentioned earlier the centriole is composed of microtubule filaments made from alternating copies of the proteins alpha- and beta-tubulin (see figure 9.1). These proteins are however not arranged in a random way, but according to a precise numeric organization that seems to have been unchanged for at least a billion years. At the core of this centriole structure lies an axoneme (in Greek, *axn* means "axis" and *nma* means "thread") consisting of one central microtubule surrounded by nine connected microtubules, which are organized in a ring. Usually these nine microtubules in the ring are tripled and sometimes they are doubled, which is then referred to as 9 (+2 or +1). Based on this core we can conclude that the centriole is unique as an organelle, and not just because it is straight and made up of two perpendicular components. It is also unique because its structure is subordinated to a well-defined numeric organization,* which seems to have been conserved for a very long time. It can clearly be said that no other cellular component has such a strictly defined geometry reflective of a mathematical organization.

Guenter Albrecht-Buehler, who has studied this geometry in depth and has proposed an interesting theory about cellular intelligence in which the centriole plays the role of an eye,[1] said in 1993: "It is a major shortcoming of the presented arguments that they cannot account for the ubiquitous nine fold symmetry of centrioles and basal bodies. . . . One may seek the explanation elsewhere, for example, in the reduction of thermal noise or in an accident of evolution. However the final explanation of the structure of centrioles and basal bodies must give more convincing reasons than these for the conservation of the nine fold symmetry."[2]

By themselves, the microtubules also have a strictly defined numeric arrangement, since they are invariably formed by thirteen filaments made from dimers of alpha- and beta-tubulin proteins. We have once again run into the number 13, which apparently has been conserved in the architecture of microtubules throughout the evolution of multicellular organisms. For the reader of this book, or any person knowledgeable in the Mayan calendar, the conservation of this

---

*Flagella and cilia, derived from basal bodies/centrioles, share the same numbering system, but do not have the same perpendicular structure.

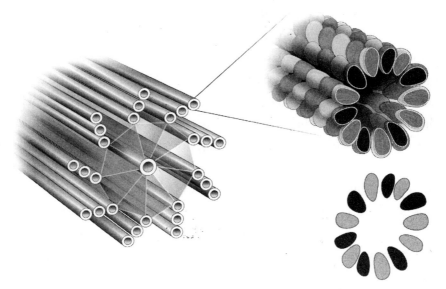

Figure 9.1. The architecture of centrioles. Schematic cross section of a centriole showing the ninefold symmetry of the microtubules, which are triplicated, 9 (+2). The component microtubules consist of thirteen filaments. A cross section of a filament shows a ring of alternating alpha- and beta-tubulins. Courtesy of Bengt Sundin.

numeric organization throughout the era of multicellular evolution will not come as a surprise. The reason for the conservation of this numeric organization is that *in animals the centriole is the primary manifestation of the Tree of Life on the cellular level.* In its design, the centriole embodies the numerology of the Mayan calendar system, nine Underworlds each made up of thirteen Heavens, describing the entire creation scheme emanating from the Cosmic Tree of Life. A line from the *Book of Chilam Balam* of the Maya, describing the creation of the world, defines the particular relationship between them: *The Bolon-ti-ku* (Lords of the Nine Underworlds) *seized the Oxlahun-ti-ku* (Lords of the Thirteen Heavens).[3] This is indeed what is evident in the centriole, where nine microtubules are each made up of thirteen tubulin filaments that they have "seized."

The point to realize is that the centriole, the Cellular Tree of Life, embodies and reproduces the cosmic plan. That this numeric organization has been conserved throughout evolution tells us that the time plan of the Mayan calendar has been there all the time and that the fundamental nature of this plan has not been altered during the course of evolution.* To make the comparison of centriole architecture with the creation of the world even more profound, we may also note that a cross section of the thirteen filaments of one of these microtubules shows an alternating ring of alpha- and beta-tubulin proteins, reproducing the rhythm of seven Days and six Nights of divine creation (see figure 9.1). It would in fact probably be impossible to create a Cellular Tree of Life that in its architecture more clearly transmits the basic structure of creation than the centriole. We may note that the same structure of nine and thirteen exists also in the cilia and flagella, and that it is from the movement of these structures in simple eukaryotic cells that the first indication of consciousness emerges. As I have argued previously, the evolution of consciousness is based on the evolution of the nine Underworlds and thirteen Heavens.

Based on this astounding numerical structure, together with the preceding discussion of organismic wave forms and centrioles, we may conclude that the design of the centriole in animals reflects the basic structure of the Mayan calendar. The numerical correspondences with the Mayan calendar are so striking that I feel we can be absolutely certain that centrioles indeed reflect the inherent rhythm of evolutionary creation and that it is the role of centrioles to mediate. It is through such a recognition—

---

*Some today think that the Mayan calendar is related to the precessional cycle of the Earth of 26,000 years, which is well-known from European astrology, and so argue that the end date of the Mayan calendar is December 21, 2012. This idea is based on a notion of mechanical time and not on quantized Kairos. In fact among the thousands of inscriptions from the Mayan sites, the 26,000-year cycle is not found once. The completion date of *the calendar must, if it is based on nine Underworlds and thirteen Heavens, end on a date that is 13 Ahau.* The October 28, 2011, date meets this requirement, but not the December 21, 2012, date. (For a view of contemporary Mayans on the end date, you may listen to my interview with Don Alejandro Oxlaj, head of the council of elders of the Maya, available from http://mayanmajix.com/dvd_nd.html.)

that the Tree of Life exists also on a cellular level and expresses the inherent structure of creation—that all of biology may logically be unified into a whole. Nothing in biology makes sense except for in light of the Mayan calendar. That the centriole reflects the basic structure of the Mayan calendar is such an amazing fact that I encourage every reader to verify this for him- or herself in textbooks or on the Internet. Given that the tubulin proteins that make up the thirteen filaments of the microtubules of the centriole are made from twenty amino acids, it can also be said that the centriole is a reflection of the basic structure of the Sacred Calendar, which consists of 260 days ($13 \times 20 = 260$). The reason why the number of amino acids is the same as the number of glyphs in the Sacred Calendar remains unknown, but there is little reason to doubt that this number also ultimately will find its origin in the Cosmic Tree of Life.

Even though the true role of the centriole has remained unknown since it was discovered more than a hundred years ago, this structure may now gain its rightful central place as the crucial creative factor in the biological evolution of animals. This evolution may now be seen as vastly more encompassing than in previous theories and is in fact only understandable as part of its cosmic context. Far from being random, biological evolution is very deterministic and directed, and we must assume that this is the reason that all cells of all animals hold the basic scheme of creation in them—nine Underworlds each made up from thirteen Heavens. There is no longer any reason to be surprised by the strict adherence of the rhythms of biological evolution to the periods of the Mayan calendar. In biological evolution the centriole serves as an antenna and executor on the cellular level for the morphogenetic fields emitted by the Cosmic Tree of Life, Hunab-Ku, whose discovery was reported in chapter 1. The overall program for the evolution of conscious animals has been provided by the centrioles, which were implanted in life on Earth at least as early as the beginning of the Mammalian Underworld. Although evolution's organismic manifestations may be somewhat modified by local environmental conditions, it has been working toward its overall aim of creating a being, the human being, in the image of the Cosmic Tree of Life for all this time. Whether we like it or not, we have been created in the image of the Cosmic Tree of Life by a purposeful universe.

Exactly how the organismic wave form unifies with the Cosmic Tree of Life or is transmitted from the Cosmic Tree of Life to the centriole is still somewhat unclear, but I will make a suggestion of a solution. The problem then is in discovering how the higher levels of the Tree of Life combine to create the wave movement to transmit the Halographic information necessary for a centriole to define the anatomy and morphology of a multicellular organism. I will here just make a few comments about it. It is currently believed that the ecliptic runs parallel with the great Central Axis through the universe, which we have here recognized as the Cosmic Tree of Life.[4] If this is an exact alignment (which is not known at the current time), the polar axis of our Sun would be perpendicular to the Cosmic Tree of Life and the two would together form a cross that may serve as a transmitter, or relay, of Halographic wave forms. Possibly then the cross formed by these axes, in combination with the Earth's own axis, may contribute to an interference pattern akin to holography that provides the information of a specific organismic wave form to be executed by the centriole on the cellular level.

I also strongly believe, without any evidence, that the Earth's own Tree of Life must play a decisive role as a Halographic source or relay in the emergence and evolution of multicellular organisms. There are several reasons why I believe that the Earth's own axis plays a crucial role as a last focusing lens for the transmission of wave forms from the Cosmic Tree of Life to the multicellular organisms. First, the microtubules and the centriole did not emerge until the Earth's inner core had solidified, which happened about 2.5 billion years ago.[5] Second, the default period of a cell cycle in higher animals is 24 hours (although sometimes it is considerably shorter, and some cells do not divide at all), which indicates a link between the cell division cycle of such organisms and the Earth's rotation rate. Third, the tubulins in the microtubules are arranged in a hexagonal structure, meaning that the Halographic information may be mediated by the Earth's inner core, whose iron crystals mostly have a hexagonal form.[6] Despite these arguments, I feel that the exact nature of the mechanism for transmitting Halographic information to the centriole at this point must be considered unknown and that advanced mathematical models must be developed to investigate this.

It should be added here that Stuart Hameroff's work with microtubules[7] has shown that they may serve as information transmitters because of the propagation of patterns of conformational changes in the tubulins.[8] This may be an important aspect of how Halographic information that has been received by the centriole from the organismic Halo may be transmitted to other cells. Such a communication system could have been developed to bring about identical transformations of cell types in mirror positions along the left-right midline of an animal. Hameroff, an anesthesiologist, has demonstrated that unconsciousness may result when anesthetizing gases block the information flow in the microtubules. From the perspective of the theory developed here, such unconsciousness would be explained by the blocking of the resonance of the organism with the Cosmic Tree of Life, the source of consciousness.

There still remain some unanswered questions regarding how biological evolution is created, and many details would require additional research to be worked out. The Tree of Life theory is not a closed system that precludes further research or discussions. Yet in terms of providing a basic framework for evolution, I feel the alpha has been brought to the omega, and the remaining problems regarding the mechanism of biological evolution are minor by comparison. Thus a study that in its first chapter started with the discovery of the Tree of Life on the cosmic level is here in its ninth chapter completed with the recognition of the Tree of Life on the cellular level. At the center of all creation is the Tree of Life, the fundamental space-time organizer and fine-tuner of this universe, a universe that is continuously being created with the purpose of generating life in accordance with the rhythm of the Mayan calendar.

It can also clearly be affirmed that this journey in eight steps—from the universal level to the cellular—has been consistently fueled by empirical evidence and logical reasoning in line with the ideal of rational empiricism introduced by René Descartes some four hundred years ago. Hence although the Tree of Life theory may be controversial because of its metaphysical implications, the factual material upon which it is based is not in any sense controversial.

The presentation of the theory could end here, so that I would leave it up to the reader to decide whether the theory is simpler and more encompassing than Darwinism. Yet I will provide some further verifications of the Tree of Life theory in order to allow us to ponder its broader existential consequences. Before I continue, I feel that it is important to point out that the centriole is not the Tree of Life, but rather a manifestation of it. The Tree of Life is primary to any of its possible material manifestations and is thus of a nonphysical Platonic nature. It is a projection of a True Cross beyond all the forms in which it has been worshipped by human beings. Yet the discovery that the centriole is a direct physical manifestation of the Tree of Life does allow us a good inroad to learning about the nature of this.

THE ORIGINS OF BEAUTY IN SACRED GEOMETRY

Although the finding that the centriole in its design directly reflects the Mayan calendar may seem astonishing at first, this is only because our whole culture is imbued with the Darwinist idea that evolution is random and undirected and not precisely timed. If however evolution at large follows a time plan, it also makes sense that this plan is integrated in our bodies and in our cells. On a macro scale it was shown in chapter 4 that evolution adheres to the Mayan calendar, and we can now understand how the pulses that this describes on a micro scale are transmitted to our cells by the Cellular Tree of Life, the centriole. The Trees of Life on different levels were earlier demonstrated to be the fundamental space-time organizers of the universe. Thus we have reason to expect that the centriole in its design reflects not only sacred time but also sacred space. The study of sacred space is part of a discipline called sacred geometry, and similar to the Sacred Calendar, its exploration may be brought to any level of complexity. Like the Mayan calendar, sacred geometry is sometimes trivialized. (The Mayan calendar is sometimes misconstrued to be about the precessional cycle known from European astrology rather than about the nine Underworlds and the thirteen Heavens.) Yet, we now have direct knowledge from the Cellular Tree of Life as to what are real expressions of sacred time and sacred space. The cosmos may speak directly to us through the way the Cellular Tree of Life has been designed.

As we study the Cellular Tree of Life in the form of the centriole, we find that it expresses sacred space and sacred time in equally astounding ways. At this point it may be pertinent to repeat that in this new theory of biological evolution, the human being has a very special place, as this is the organism that creation has been directed toward all along. This means that we should expect ourselves to a higher degree than any other species to reflect the Cosmic Tree of Life in its capacity as an organizer of both time and of space. Thus we, unlike most other species, have an estimated 260 cell types (different estimates exist however), which happens to be the number of days in the Sacred Calendar. As another example, the average time of a human pregnancy is 266.2 days,* which is closer to 260 than that of most, or all, other species.

If we now shift our focus to sacred geometry, we may first note that *the centriole embodies* two geometrical structures that are basic to the creation of the universe, namely *the straight line and the circle,* which are rare, if not unique phenomena to be found in biological matter. The combination of these two may by itself account for the creation of many, if not all, forms. Second, the tubulin proteins on the microtubule surface are organized in a hexagonal pattern. This hexagonal pattern,

---

*According to the Nägeles Rule, named from a German doctor at a hospital in Heidelberg

present also in snowflakes,* is the starting point for a branch of sacred geometry based on the so-called Flower of Life,[9] which was explored by Leonardo da Vinci. What is noteworthy in this context is that the tubulin structures present on the centriole are known as the Seed of Life and the Fruit of Life (figure 9.2). The Fruit of Life has been said to hold a blueprint for the universe, and if we assume that the Earth's inner core with its hexagonal iron crystals plays a role in the Halographic projections transmitted to the Cellular Tree of Life, there may be some truth in this. The point to realize is that in the centriole, we may directly study the Sacred Geometry of the *real* Tree of Life on a cellular level and its expression in biology. Such studies may allow us to verify what previously might have seemed like abstract esoteric ideas. We now have tangible reasons to believe that certain geometric structures, if they are present in the geometric architecture of the centriole, may reflect principles and patterns that are of *primary* importance for the creation of ourselves and the universe, and the centriole may thus serve as a meaningful guide for the practice of sacred geometry.

The golden ratio, which was defined in chapter 5, figures prominently principally in the sacred geometry literature for three reasons. First, it is extensively found in nature, ranging from spirals in nautilus shells to galaxies. The Fibonacci numbers are also directly expressed in many ways in nature, maybe especially in plants. This is called phyllotaxis (the arrangement of leaves on a stem) and is expressed for instance in the number of petals in flowers, the angle between rose petals, spiraling seed patterns and needles on cacti. Maybe such structures are direct effects of the corkscrewing that characterizes the Cosmic Tree of Life.

Second, the golden ratio is considered especially pleasing to the human eye and to be a norm of beauty in nature as well as in the fine arts.[10] The golden ratio is a proportion that has been used, consciously or unconsciously, by artists and architects possibly as early as in the Egyptian pyramids, but more certainly in the Parthenon temple as well as in the

---

*Maybe this is what is behind the effects that Dr. Emoto sees in snowflakes.

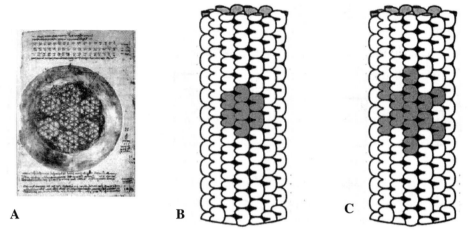

Figure 9.2. The Flower of Life geometry in the Cellular Tree of Life. The Flower of Life geometry as drawn in a manuscript by Leonardo da Vinci (a) is evident in the structure of tubulin proteins in the centrioles as the Seed of Life (b) and the Fruit of Life (c).

great cathedrals of medieval Europe. Many works by Leonardo da Vinci are well known for having incorporated the golden mean in their proportions, and this may be part of the reason why they have been considered beautiful. Interest in sacred geometry may in fact have increased with Dan Brown's book *The Da Vinci Code* and its highlighting of mathematical relationships in nature, sometimes supposedly kept secret.

The third reason for the interest in the golden ratio is that it is at the center of several interesting mathematical relationships. A whole philosophy has been developed around the golden ratio, with its many expressions in nature seen as the works of the Creator. A well-known religious expression of the golden ratio depicts the Hindu god Shiva, the creator and destroyer, holding a nautilus in one of his hands, whose shell is known to reflect the golden ratio.

Exactly why this ratio manifests so abundantly in the universe has been a mystery all this time. I feel that we, however, on our journey to understand biological evolution, have now stumbled upon the solution to this enigma. To discover the origin of the presence of the golden

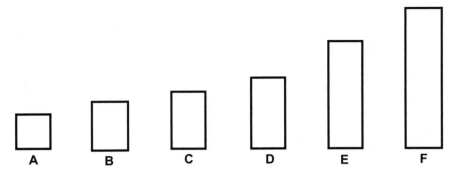

Figure 9.3. The aesthetic qualities of different rectangles. The German psychologist Gustav Fechner studied which proportions in rectangles seemed the most pleasing to human observers. He asked them to pick the one in a row they liked the best. The reader may pick the rectangle he or she finds the most aesthetically pleasing and compare the choice with that of others; see footnote below.

ratio in nature, we must first find out if there is any basis to the claim that it is especially pleasing to the human eye. To investigate this, a German psychologist by the name of Gustav Fechner performed an experiment where he arranged a series of rectangles, as in figure 9.3, and asked various subjects which rectangle they found to be the most pleasing aesthetically. The reader of this book may wish to note which rectangle in this row he or she finds the most pleasing and compare this to how others have responded.* It turns out that most people indeed prefer the rectangle that is an expression of the golden ratio, which is also called a golden rectangle. While tastes obviously vary among human subjects, this would indicate that our experience of a pleasing nature of the golden ratio was somehow built into us from the start.

If we now return to the centriole, and the picture of it taken by electron microscopy in figure 8.6a, an interesting fact is revealed to us.

---

*About 35 percent of those asked favored the rectangle with the proportions in C, about 20 percent the two neighboring rectangles, while the other ones were much more rarely selected.

In this figure we can see that its proportions conform exactly to the rectangle (figure 8.6c) that is shown to the side of it. The proportions of the rectangle in figure 8.6c is identical to the golden rectangle shown in figure 9.3 (item C). This, in other words, means that the centriole is designed according to the proportions of a golden rectangle. What this concordance means is that given that we have found that the centriole, the Cellular Tree of Life, plays the predominant role in defining the body plans of multicellular organisms, we would expect to find the golden ratio and the golden rectangle expressed in their anatomies. This is indeed exactly what we find, and especially so in the human being (figure 9.4), which is the species that we have found to reflect the Tree of Life to the highest degree. Thus expressions of the golden ratio found in animals ultimately find their origin in the geometry of the Cellular Tree of Life and it is no exaggeration to say that *the phi proportion of the Tree of Life provides the foundation for all reflections of the golden ratio in nature.* This provides the solution to an age-old enigma, namely why the anatomy of the human body expresses the golden ratio in so many ways. While truly astounding, once it is realized, it also seems like a perfectly logical conclusion that the Tree of Life would be at the root of all true sacred geometry, quite in line with the material we have presented so far.

In this we have actually found another verification of the new theory of biological evolution, since a critical part of this theory is that the centriole, the Cellular Tree of Life, is at the origin of the anatomy of all animals and most notably of human beings. Human beings, because we have been created by four different Underworlds, embody the golden ratio of the Tree of Life to a higher degree than any other species, as evidenced by the simple fact that our faces essentially are golden rectangles. Another case in point is that, in human beings, the ratio of the

---

*This would include the microtubule structures from which centrioles are made and the cilia and flagella that are derived from them. In plants, lacking centrioles, it is probably the microtubules that cause the so-called Fibonacci phyllotaxis.

distance between the feet and the navel to the distance between the feet and the top of the head is equal to phi (see figure 9.5). This means that the divine proportion was expressed through the centriole from the point when the embryo first attached to the wall of the uterus (from which point the navel later developed). Interestingly, however, and quite consistent with the theory that it is the Halo of the *adult* organism that drives development, the ϕ ratio does not manifest in the newborn baby (where the navel is about halfway from the feet to the head), but only as the person approaches adulthood.

A fractal development of the golden ratio is also evident in figure 9.4. This indicates that the anatomical expressions of the golden ratio unfold in a specific sequential order that in principle should be possible to track. Other examples of phi in human anatomy include: (1) the human head forms a golden rectangle; (2) the mouth and nose are each placed at golden sections of the distance between the eyes and the bottom of the chin; (3) the ratio of the whole body height to the distance between the head and the fingertips equals phi; (4) the ratio of the distance between the top of the head and the fingertips to the distance between the head and the navel or elbows equals phi; (5) the ratio of the distance from the top of the head to the navel or elbows to the distance between the head and the pectorals and the inside top of the arms equals phi; (6) the ratio of the distance from the top of the head to the navel or elbows to the width of the shoulders equals phi; (7) the ratio of the distance from the top of the head to the navel or elbows to the length of the forearm equals phi; (8) the ratio of the distance from the top of the head to the navel or elbows to the length of shinbone equals phi; (9) the ratio of the distance from the top of the head to the pectorals to the distance from the top of the head to the base of skull equals phi; (10) the ratio of the distance from the top of the head to the pectorals to the width of the abdomen equals phi; (11) the ratio of the length of the forearm to the length of the hand equals phi.

The many findings of the golden ratio in human anatomy are direct expressions of the golden rectangle of the Cellular Tree of Life, the cen-

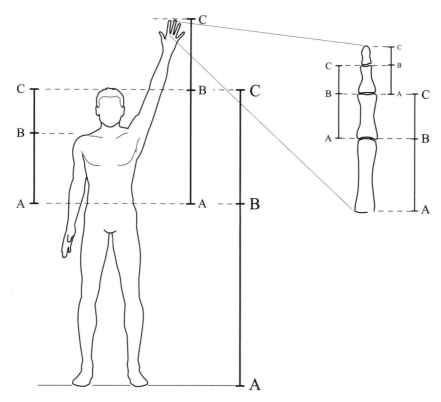

Figure 9.4. The golden ratio and human anatomy. The golden ratio is evident in many proportions in the anatomy of the human body. Courtesy of Bengt Sundin.

triole, according to the Tree of Life theory. We may also note that many ratios involve the navel, from which human anatomy is developed when the organismic wave form starts to be expressed by the primordial centriole in the fertilized egg. While the relationships among embryonic development, centriolar positioning, and human anatomy still need to be worked out in detail, I do not think that there is any reason to doubt that such relationships indeed exist and are precisely determined by the organismic wave form.

To distinguish the current work from the many books that have previously been written about the golden ratio, I would like to point out that in this theory, the centriole, or the Tree of Life at any other level, is not just another example of how the divine proportion is expressed in nature. Instead, *the Tree of Life is the very origin of the divine proportion in nature* and through the centriole this relationship is being revealed to human beings for the first time. We may also surmise that it is the golden ratio's origin in the geometry of the Cosmic Tree of Life, the source of our existence, that makes human beings perceive the golden ratio as both divine and pleasing. Thus the golden ratio is a geometry both at the source of creation and at the root of our being.

We have found that the Tree of Life is the fundamental organizer of space, not only on the cosmic level, but also on the organismic and cellular levels. There it creates organisms much like how pre-Darwinian biologists envisioned anatomy and morphology to be the results of "crystallizations" of cells in accordance with certain Platonic forms, sometimes according to the ideals of sacred geometry. There is no reason to be surprised that the golden ratio through these archetypes plays a crucial role in defining the relationships between organisms and cells, since the golden ratio indeed is a mathematical expression for a harmonious ratio between the whole and the parts. If to this we add that the centriole embodies the straight line and the circle, its power to create a variety of forms should be clear to everyone. (The constants phi and pi thus associated with the Tree of Life are arguably the most significant constants of geometry.) Through this theory it has thus been possible to create a bridge from sacred time and sacred space, which have long been studied by ancient cultures and mystics, to the source of creation. An extensive connection has also been established between the Platonic world of "ideals" and their physical expressions as biological species.

What we are now starting to discover is that the anatomy and morphology of biological organisms are not the mere results of survival of the most functional traits, such as Darwinism postulates. Instead, they are created based on what we may call different quantized Platonic archetypes, which are at the root not only of the beauty of these organisms, but also of their intelligence and compassion. Because of the Sacred Geometry of these archetypes we may understand that to many people, nature is very beautiful and the source of spiritual experiences. Nature is not merely functional, but it, and all of its species, has a soulful essence. We are here returning to the pre-Darwinian concepts for the emergence of the biological species, such as those expressed by the German writer and scientist Johann Wolfgang von Goethe: "We are not to explain the tusks of the Babirussa by their possible use, but we must first ask how it comes to have tusks. In the same way we must not suppose that a bull has horns in order to gore, but we must investigate the process by which it comes to have horns in the first place."[11] In Goethe's view, form is a primary and function a secondary adaptation, and I believe the reason so many of us are able to experience beauty in nature is that nature is not designed primarily based on functionality, but on forms that echo an underlying harmonious geometry.

In chapter 4 we saw that the three-dimensional anatomical form indeed is the primary expression of evolution, meaning that physiological and biochemical adaptations are secondary to this. In chapter 6 we went on to argue that the Mayan calendar, which this evolution follows, is quantized. This means that an animal or a plant cannot look in any possible way just as long as it is functional. Because the pulses emitted from the Cosmic Tree of Life are quantized, each species is created by a unique and distinct Platonic archetype, or wave form if you like, which expresses its very essence. Similarly to how atoms may take the form of either oxygen or nitrogen, but never a mixture of the two, a biological organism has a distinct quantized character. An animal may thus be either a babirusa or an elephant, but there is not a continuum of species between the two. An animal or plant cannot take just any form, but only one that corresponds to quantum states of the Cosmic Tree of Life. This is the reason that each animal has a unique essence corresponding to its Platonic archetype. This idea of an essence would clearly be more consistent with ancient views, such as the biblical ("created after *their kind*"), but it would also be consistent with those of shamans of indigenous peoples, who tend to look at animals as naguals, or spiritual totems, ensouling each particular animal or plant species.

Because the quantum numbers of the Mayan calendar are limited, there are only a limited number of discrete Platonic archetypes that may define the characteristics of a biological organism. If we look at the totality of multicellular organisms, the number of such archetypes may well range in the millions, each generated by a particular quantum state of the Cosmic Tree of Life. Yet the number of such different polarizations would not be endless, and the species generated by them would all be different in their essences. In the case of members of *Homo sapiens,* which is the only surviving species of the Human Underworld, it is also clear that they are the result of a unique quantum state of the Cosmic Tree of Life (Mayan calendar quantum numbers 12.12.12.12, plus some additional minor polarizations), even though within certain limits genetic and epigenetic factors may create a variation in how the Platonic archetype of human beings is expressed. Taken together with our earlier findings of the many anatomical expressions of the golden ratio in human beings, this elevated quantized Platonic form may give a more tangible meaning to the statement in Genesis that human beings have been created "in the image of God." In addition to the geometry of the human being, the golden ratio defines the geometries of the atoms of life (C, N, O, P, S), the molecule of life (DNA), and the organelle of life (the centriole) creating expressions of Sacred Geometry on each of these levels. If we ever wanted proof that a Platonic geometric form, the golden ratio, characterizes all life, we have it here (see figure 9.5).

In my own mind the discovery of the golden ratio, especially in the centriole, but also in the general characteristics of life emanating from the Cosmic Tree of Life, may be the most beautiful verification of this theory of biological evolution. As a personal aside I should say that I have not previously taken any interest in sacred geometry, and I actually happened to more or less stumble on the discovery that the centriole is a golden rectangle as a bonus after the theory had already been fully developed in the way described in previous chapters. Two good friends, Phil den Ouden from Hertme in the Netherlands and Les Tulloch in Vancouver, British Columbia, had however given me books on sacred

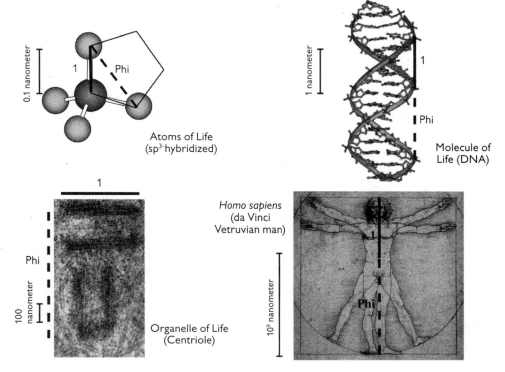

Atoms of Life
(sp³-hybridized)

Molecule of
Life (DNA)

Homo sapiens
(da Vinci
Vetruvian man)

Organelle of Life
(Centriole)

Fig 9.5. Manifestations of the golden ratio at different levels in the hierarchy of life.

geometry the year before I started on this book, which led me to look more closely at the geometry of the centriole. To me, this added verification that the Tree of Life theory is not just a personal musing about biology and physics, but a theory that is objectively true.

IS GOD A MATHEMATICIAN?

What is done by what is called myself is, I feel, done by something greater than myself in me.

JAMES CLERK MAXWELL

After having seen some of the expressions of sacred geometry in creation we may now want to go back somewhat and ask about the origin of numbers. Not surprisingly, I suggest that their origin is to be found in the Cosmic Tree of Life, particularly in the duality that this creates, since our two hands, two breasts, two feet, and so on seem to be the most likely origin of counting, as they provide a basis for taking the step from "one" to "two." If we think about it, if such a duality was not present in the anatomy of animals or human beings, would there ever be anything but "one" or "many"? Would the phenomenon of counting, taking the step from one to two, ever have emerged without the existence of this basic duality? That two-ness, or duality, is a very special step in counting is demonstrated in Hebrew, which has a special form of plural for identical pairs, such as eyes, feet, and pants. In some languages, counting is based exclusively on the numbers one and two, so that one, two, three, four, and so on are counted as one, two, two-one, two-two, and so forth. This seems to indicate that it is the basic duality of the Cosmic Tree of Life and its reflection in anatomy that may be at the origin not only of the concept of "two-ness" and the number 2 but also of the very phenomenon of counting. The fact that many counting systems have their bases in the numbers 5, 10, and 20 seems to support the idea that counting has its origin in bilateral anatomical features such as fingers and toes.

In the prehistory of humankind, counting presumably was the first step toward mathematics. As the mental ability to isolate concepts and elaborate them abstractly was developed in the higher Underworlds, so did methods for computing, such as addition. From addition, the most basic form of computation, subtraction, multiplication, and division were derived, and from the abstract concept of number, mathematics moved on to much more advanced ideas. But if all of this ultimately goes back to the number 2, and this has its origin in the perception of bilateral anatomy, this means that all of mathematics has its origin in the Cosmic Tree of Life, which is the source of this. Ultimately, the reason the human may be a *Homo mathematicus* is that we are created in the image of the duality of the Cosmic Tree of Life and so more than any other species are able to reflect and describe the inherent mathematical design of the universe. This then literally also means that our "intelligence," or at least its mathematical side, is a reflection of the Cosmic Tree of Life, a very consequential statement about who we truly are as human beings. It provides the answer to an age-old question that in different forms has been asked many times before, from Pythagoras onward, namely if God is a mathematician or, in other words, if the universe has been created primarily in accordance with mathematical principles.

What we have found is that the basic concepts of counting and numbers may derive directly from the way that the universe was created by the Cosmic Tree of Life. This means that the universe could not have been created in its present form without mathematical principles at its foundation, as we saw in chapter 5. We are now however getting a sense of why we may be able to grasp these principles. If we ourselves were not created in the image of the Cosmic Tree of Life, we would never have been able to begin to grasp even the basic concepts of mathematics or how they might apply to the creation of the universe. I am now returning to a theme I hinted at early on, namely that it is because of our left-right polarity that we are intelligent. The need to create intelligence through duality was the primary reason that evolution took the step from the unicellular life forms in the Cellular Underworld to the multicellular organisms in the Mammalian. Without this step to multicellularity, the duality of the bilateral biological organisms would never have emerged and as a consequence no intelligent beings reflecting the Cosmic Tree of Life would ever have emerged. Thus only bilateral animals are intelligent, and we can understand why the Mammalian Underworld had the creation of intelligent beings as its purpose.

It is because our own duality is a reflection of the duality of the Cosmic Tree of Life that we are intelligent to begin with. Certainly, reducing intelligence to mathematical abilities oversimplifies things, but I feel that counting is an expression of a more basic human ability, namely that of isolating mental concepts and elaborating them abstractly, and so it is likely to reflect a significant step toward the emergence of intelligence in a much broader sense. Regardless, the conclusion is that if a Cosmic Tree of Life did not exist, and if we had not been created in its image, our modern advanced technological civilization, which in fundamental ways depends on our mastery of mathematics, would never have developed. I feel this is a very profound insight, indicating that intelligence is a function of resonance with the waves emanating from the Cosmic Tree of Life and not something that has its origin in brain chemistry. (This view incidentally is fully consistent with Karl Pribram's Holographic model of the brain.) I also feel it is a reason for concern that this origin of our intelligence, although it was recognized by the ancients, has been denied throughout the era of technological development. What direction does a civilization take when it is all built on denying its origin? Most likely it goes in the direction of uncontrolled technological and economic growth without concern for the purpose that the universe was originally designed for, a direction that incidentally now seems to have backlashed. The origin of our intelligence in the Cosmic Tree of Life is also generally a cause for us human beings to be humble in the realization that all we really are able to do are things that ultimately are reflections of this awesome power of creation.

We have already seen some direct examples of how the universe has been created in accordance with mathematical principles. As mentioned in chapter 1, the Cosmic Tree of Life seems to be the origin of the three dimensions of space, and it is then no stretch to consider that it may also be the root of all geometry. Through its geometry, the centriole may also be considered to be at the root of the critical numbers 2, phi, and pi. Thus mathematics is not only geometry with its lines and figures, but also arithmetic, algebra, and analysis leading into higher mathematics. Mathematicians and physicists have at the occasions of great discoveries often felt that their equations had an existence that seems independent of themselves, and so have given them a glimpse of a higher reality. What may have struck them like a lightning bolt is that certain equations or mathematical relationships somehow seem to have existed in a higher domain, where they were only waiting, or even prompted, to be discovered. When this happened, it may have created an experience of euphoria or transcendence in the discoverers, which is expressed in the quote at the beginning of this section by James Clerk Maxwell, the father of the theory of electromagnetism. Mathematicians and physicists sometimes express that certain equations or theories are beautiful and originate in a higher domain. Indeed, if the Cosmic Tree of Life is the primary space-time organizer of the universe, and at the same time at the root of our sense of beauty, then there is no reason to be surprised that the discoveries of its creative principles in mathematical form give rise to a spiritual experience of beauty on the part of the discoverers.

It should be clear that mathematical relationships are not mere human inventions, but are relationships embedded in the Platonic reality emanating from the Cosmic Tree of Life. Our material world, where physicists discover mathematical relationships and profound equations, is ultimately a reflection of this higher reality. In fact, material reality is only a reflection of, and secondary to, the world of sacred mathematics, which has always been the view of Pythagoreans. Very often, those scientists who have been open to a view of a divine origin of the universe, such as Kepler and Einstein, have been the discoverers of the most beautiful theories of science. Many thinkers have for a long time embraced this overall Platonic view, but now, with the tangible discoveries of the Tree of Life on both the cosmic and cellular levels, its truth will hopefully become much more evident. A Platonic "world of ideas" really does exist in the form of the Cosmic Tree of Life and the vibrations it creates, and certain equations and mathematical relationships are expressions of this.

I am proposing that the Cosmic Tree of Life may be at the root of beauty not only in aesthetics but also in mathematics. Because of its mathematical nature it is also known as the *Tree of Knowledge,* and in ancient traditions it was often referred to as a source of wisdom. I believe the reason we perceive the golden ratio as beautiful, whether in art or mathematics, is that as an aspect of the Cosmic Tree of Life, it reflects the very source of our existence. The ramifications of the discovery that the golden ratio and several other aspects of sacred geometry and sacred mathematics are rooted in the reality of the Cosmic Tree of Life are so far reaching and inviting to further exploration that they would easily generate several books. Also, the fact that the Tree of Life exists in every cell of our bodies and is at the root not only of biological, but also of mental evolution, has so many consequences that I would rather stop here and leave this to the reader to explore. What is presented here I believe is only the critical opening of the portal of the Tree of Life.

In a broader sense I think we now have every reason to recognize that phi, the golden ratio, is as much a constant of our particular universe as is the tun. Unfortunately, we do not know if the golden ratio dominates the cosmos at its largest scale, since we cannot observe the whole universe, but only a sphere with a radius of some 15 billion light-years. Yet one of the explanations that cosmologists have proposed for the origin of the Central Axis is that the universe has a different extension in different directions. Hence, I will assume that the Cosmic Tree of Life has axes whose sizes conform to phi and that the centriole is a reflection of this. If the universe as a whole is dominated by the golden ratio constant, this would mean that figure 1.4, showing equal axes, would not be strictly correct. Since the centriole embodies both the golden ratio and the circle, we may also guess that those two geometric constructs are the primary ones from which our universe has been designed, something that was evident in our study of the electromagnetic interaction.

This raises the question of whether other universes do exist and to what extent they may have been designed in a way that is different from ours. We have seen some advantages and consequences of designing the universe based on the tun system and the golden ratio, but does this mean that these definitions of time and space are the only ones possible for a living universe? Would it be possible to alter some of the other constants of nature and yet create a universe that succeeds in generating life, but maybe in a different way? Consider for instance the possibility of a universe where the basic dimensions of the Cosmic Tree of Life would not be defined by the golden ratio, but by a square. This conceivably would lead to a cousin to the human beings in another universe where all the anatomical ratios listed in the previous section would be equal to 1, rather than to phi, including that of the face, which would then have a perfectly square appearance. Would such a universe be possible? Personally I doubt it and think that it would be too symmetrical for life to thrive, but this is merely a guess. Regarding the bigger question of the possible existence of different universes, I hope that there is now somewhat more of a foundation if we want to play God and design a universe. Maybe we will find that even though most of us at one point or another are saddened by the amount of suffering in the world, or in our own lives, and some even argue that this precludes the existence of a Creator, it is not necessarily an easy thing to create a universe with a hundred billion galaxies where every single being is happy all the time.

TOWARD A UNIFIED MEDICINE

The development of a new science of biology would be expected to result in the emergence of a new medicine, since after all biology provides the scientific foundation for the practice of medicine. Here a sound scientific theory has verified that the wholeness of our bodies has its origin in the same sacred domain of reality as is accessed by artists and mathematicians as they have experiences of the divine, namely that of the Tree of Life. This is a new (and old) view of biology, and if the theory becomes known and accepted more broadly, it is bound to have significant consequences for medicine.

It seems however that a shift in the direction of medicine to a large extent is already taking place. As so often happens, the changing attitudes and practices of people at large precede the formulation of new scientific theories, such as the one presented in this book. In past decades many Eastern medical practices have become accepted or even mainstream in the West, and in the United States there are now as

286   A Biology with Soul

many client visits to practitioners of alternative or complementary medicine as there are to regular doctors.[12] I mention Eastern practices here as an example of a shifting direction of medicine, as these are usually more in line with the holistic concept that our bodies, and really our whole beings, are the result of a purposeful integration brought about by organismic Halos. Thus while broadly speaking Western medicine is organotropic, meaning that it seeks to treat diseases essentially only in organs in which there are symptoms, Eastern medicine is holistic and seeks to balance or harmonize the energy system of the whole body when healing a localized condition.

Yet holistic practitioners have often had to fight an uphill battle against the medical establishment. The reason for this has partly been that the new practices have not always been consistent with Darwinism and its theories about living organisms that has dominated academic biology. As is often pointed out in textbooks of biology, Darwin's theory of biological evolution provides the fundament of all of established biology, and its influence has reached very deep into many aspects of medicine. The new approach to biology in this book is however essentially holistic in that it has been demonstrated that what primarily determines the nature of our beings is a Platonic archetype and that all the physiology and biochemistry of various organs, tissues, and cells are fundamentally subordinated to this whole.

The basic assumptions of Darwinism also play a decisive role in what questions are asked and what subsequent experiments are performed in medical research. This research often has a circular way of reasoning so that the results are used to solidify Darwinist theory almost regardless of how they turn out. This, incidentally, is why in the year 2009, when the anniversaries of the birth of Darwin (1809) and the publication of his *On the Origin of Species* (1859) are celebrated, exhibits and articles in magazines provided by the scientific establishment will almost with one voice say that all the research since Darwin has confirmed his theories. The Human Genome Project provides a clear example of what happens when the thinking has to be within this box. When the results of this project, as mentioned earlier, contradicted the Neo-Darwinian dogmas, they were typically used to prompt more research based on a new twist of Darwinism rather than to carry out what would have been logical—dismissal of the whole theory. Because all academic research has to be within the Darwinist box, we know very little about the potential consequences of other biological theories, for instance within medicine, since very little research outside of the Darwinist model has ever been funded. Ben Stein's documentary *Expelled: No Intelligence Allowed* shows how criticism aimed at Darwinism is being purged from the world of established science.

I should point out that the Tree of Life theory does not imply that all existing knowledge of biology should be thrown out. This theory is based on the very same facts as the Darwinist model, although the facts are looked upon and interpreted differently. The aim is to place the knowledge of modern biology in a new context, one that integrates the old Darwinist theory into a wider and more encompassing theory about biology and the evolution of the universe. In this context the concepts of random mutation and natural selection would be assigned to considerably smaller roles than they previously had. In the new biology, roles for both the Halos of the Tree of Life and genome-based biochemistry are recognized even though the Halos and the cosmos at large are clearly playing senior roles in development and evolution. As a consequence of its inclusiveness this theory by necessity assigns valid roles both to Eastern holistic and Western biochemical medicine, but within a new context. Hence, while the introduction of the Tree of Life Halos fundamentally shifts our understanding of biology, all we can say at the present point in time is that the validity of both approaches needs to be explored in a new light so that they may be harmoniously integrated in the practice of medicine. It should be clear however that the new theory of biological evolution does tear down the walls between these different approaches to medical practice, holistic and biochemical, so that it is no longer logically possible to insist that only one of them is valid. This in itself means a tremendous shift. The Halos, the centrioles, and biochemical processes in the new theory all form part of a whole, the human being. The medical consequences of this integration remain to be explored, but it is already clear that what broadly is called holistic medicine plays a legitimate and promising role in the future of medicine. The reason is that we have now found that in biological organisms the parts are subordinated to the whole and that this goes back to the Platonic archetype. Then, of course, as with all medical practice, it is only the results in patients that may serve as a basis for assessing its value.

The potential for centriole-based medicine especially needs to be explored. We may recall that the organismic Halos are senior to interactions with energy and matter and hence outside the reach of physical manipulations. Thus at first sight they may seem inconsequential for medicine. In fact, from a medical standpoint, there would seem to be no reason for anyone to want to manipulate the Halos, since these define the knitting pattern and the body plan of the *healthy* adult, and health problems in this new perspective are to be regarded as failures of the physical organism to adhere to this plan. However, the very fact that an ideal manifestation of a Platonic archetype has been found to exist may be important for medicine, since it is something that by itself may influence our philosophy about health issues.

The concept of the Halo is of course not new. It features prominently in Christian art in icons of Jesus and saints. Auras that emerge from mental, emotional, or spiritual etheric fields that surround an individual are mentioned in esoteric literature, and the chakra system plays an important role for our being in Ayurvedic medicine. Some healing techniques, such as Therapeutic Touch, that are based on the practice of healing the Halo, have been found to have a supporting influence. What is new here is that the notion of our Halo, an idea that has a fairly ancient origin, is to be regarded primarily as an emanation of the Tree of Life. It is in other words not primarily an emanation of the individual, but of the individual's sacred origin in the Tree of Life, and it is this that truly places a person in a cosmic context. Therefore the relationships of auras to the cosmos and the human body may now be looked upon in a new way.

The organismic Halo would then, in theory at least, provide the guiding light body, or Platonic archetype, for an *ideal* state of health. It maintains the basic integrity of our bodies so that we retain the same bodies, with essentially the same appearance, throughout our grown-up lives, despite the fact that in a lifetime, due to cell division and the turnover of matter, all of our individual cells and atoms are replaced many times over. Such a light body may also generally provide an explanation for the phenomenon of healing, and it should be noted here that a theory about the healing of wounds is currently missing in established medicine. Although Western medical practitioners know very much about how wounds heal and have developed practices for facilitating such healing, there is really no room in the established paradigm to explain *why* wounds heal, partly because this would require the recognition of purposeful processes in the universe. From the existence of a Halo, providing a knitting pattern between centrioles, we may however understand why excised parts of bodies sometimes may be regenerated, and why wounds sometimes are healed. Another phenomenon that is relevant to discuss in this context is the so-called phantom limb, the sensation that a lost arm or leg, for example, is still attached to the body. The existence of a Halo, or a wave form, could explain why in such cases the experience of the integrity of the body is retained. Even more relevant with regard to the Halo are reports that people other than the subject can feel when they are touched by a phantom limb,[13] tentatively indicating that the Halo has an existence that can be experienced.

The fact that centrioles move in cells that are active in healing wounds also indicates that they somehow "know" in what cell they are located in the body and how they should relate to neighboring cells in an "ideal" state of health. I have hinted at how this "knowledge" may be possible because of the network created between mother and daughter centrioles and the spatial coordinates these provide to every cell. This would indicate that there is some kind of feedback system to the Trees of Life on higher levels or that the Halo from which the organism was formed remains in existence throughout our lifetime, either through a network of communication or some kind of light body. Interesting in this context are the findings that centrioles are receptive to light, especially infrared light.[14] If centrioles are like "eyes" and thus receptive to different frequencies of light, it is possible that a signal system using light may be maintaining a light body or even generating feedback information to Trees of Life at higher levels. If centrioles are able to serve as cellular "eyes" and recognize each other at a distance, it seems that it would be possible for them to position themselves with exact distances from one another in such a way that they would maintain the body plan of the Halo. Correctional light signals between centrioles maintaining the wave form of the body plan could thus conceivably provide an explanation as to why wounds heal. Even though Halos are beyond mechanical manipulations, it is possible that a centriole-based medicine could be developed with beneficial results. To address these issues new research seems necessary to study how the alignments of centrioles change in development and how this may be influenced by different types of signals.

With the new approach to biology, disease will be seen as a consequence of a disturbance of the expression of the "ideal," as defined by the Halo. The real question is whether there are ways we can assist the alignment of centrioles to the "ideal" defined by the Halos if the expression of this ideal has been distorted. If this were possible, it could conceivably support the healing of physical wounds, the normalization of malformations, or even balance disturbances of metabolism, which would have consequences for a large number of diseases and ailments. Along the same lines we may also ask if Halos play a role in preventing aging. These are open questions as of now, and we do not know if there are any ways we may beneficially interact with the centrioles or if there is some kind of "light body" of connections between the centrioles that we would be able to influence to help reestablish the healthy ideal. All that we know is that as the body starts to heal a wound, the centrioles move and position themselves differently in the cells[15] to help bring about the healing. Theoretically speaking, it is hard to see how such processes may be initiated if there is not some kind of light body or Halographic ideal that tells the centrioles how the wholeness of the body is to be maintained.

The seniority rule in the relationships between the organismic Halo and the centrioles is bound to have consequences for our understanding of cancer. After all, cancer is a disease in which an individual cell ceases to follow the seniority rule and starts to grow and replicate without any consideration for the health of the organismic whole. Over the years several mechanisms for this have been proposed, but it seems obvious that they are very directly related to the theory one embraces to account for the organization of cells. What should be emphasized here is that observations under the microscope have shown that in tumor cells, the centrioles almost invariably display signs of malformation,[16] and it has thus been proposed that cancer could be caused by disturbances of centriole function.[17, 18] Given the place we have here assigned to the centriole in the development of the morphology of multicellular organisms and the differentiation of their cells, such a theory seems very reasonable indeed. Yet it is also a well-established fact that agents that cause random mutations in DNA cause cancer,[19] which is known to be true for both smoking and exposure to ionizing radiation. These findings are not surprising, considering that random DNA mutations, as pointed out earlier, are able to impair the expression of the body plan and destroy the synchronization of the cell cycle. This is of great interest to me, since I have spent twenty years in laboratories aiming to develop techniques to identify cancer-causing agents in the human environment.[20, 21] I still consider this a meaningful aim, since in the new perspective, DNA-damaging agents disturb the expression of the body plan in such a way that cancer may result. Yet I no longer see DNA damage as an explanation for what cancer is, and I do not agree when some colleagues call cancer "a disease of the genes." Although damaged genes may initiate cancer, cancer in my view is ultimately a breach of synchronicity brought about through distortions in centriole function. It is really an expression of a violation of the ideal body plan that was provided by the organismic Halo that created us as mirror images of the Cosmic Tree of Life. The medical consequences of such an altered view, where disturbances in centriole function are studied in a new light, still remain to be explored.

TOWARD A REUNIFICATION
OF SCIENCE AND SPIRIT

Medicine and other fields directly influenced by biological science, such as agriculture and pharmacology, are by no means the only disciplines that over the years have been profoundly influenced by Darwinism. The Darwinist concept of "evolution," based on the struggle for survival, has extended also into linguistics, anthropology, and other fields in the humanities that found utility in applying a model from the natural sciences. To exemplify this, we may view the attempts to develop a theory about the nature of consciousness, which is something that currently attracts much attention. To explore the origin of consciousness, Darwinists have typically asked, What is the selective advantage of consciousness? To ask such a question is however to miss the whole point, since the generation of consciousness is part of the purpose of the universe and not something that just happened to pop up somewhere on the road of evolution. This illustrates how, by asking misleading questions based on a flawed and outdated theory, emerging fields of science can be led astray by Darwinism, and this is exactly what has happened several times in the past because of its influence in the humanities.

Darwinism has undeniably had a profound impact on modern life and has thus been presented as a fundament not only for biology, but also for many other branches of science. It has completely altered the view that modern people have of themselves and their place in the universe and has also had a decisive influence on their political ideologies. Eugenics, the infamous "racial hygiene," which came to play such a fatal role in history, is a case in point, proposed by Frances Galton, Darwin's cousin and a strong supporter of his ideas. While it can be argued that Galton's ideas were just reflective of the overall racism that came out of the era of European imperialism in the 1800s, I do not think that the role of misused scientific authority should be underestimated. Thus the refutation of Darwinism made here invalidates enormous amounts of speculation and philosophizing that have been presented in the past 150 years concerning notions such as "survival of the fittest" or natural selection as the driving forces behind evolution.

The effects of Darwinism however have been much more pervasive than the case of racism mentioned above, and more than any other scientific theory, Darwinism has presented a picture of the universe as spiritually dead and lacking purpose. Its predominance in the sciences is expressed by physicist Leonard Susskind: "Modern science really began with Darwin and Wallace. Unlike everyone before them, they provided explanations of our existence that completely rejected supernatural agents. . . . Darwin and Wallace set the standards not only for the life sciences but for cosmology as well."[22] Thus, Darwinism has come to dominate the science of physics, since it has cemented an approach of pointlessness and godlessness for the universe that has strongly determined its basic principles and the kind of questions that this science has been asking. The study of the fine-tuning of the constants of nature presented earlier could not have come out of a physics inquiry under the influence of Darwinism. The findings presented here about sacred geometry and sacred mathematics would have had to be relegated to metaphysics. Yet these findings are very relevant to cosmology in general and to physics in particular and maybe even more so is the discovery of the Cosmic Tree of Life as the fundamental space-time organizer of the universe, with potential consequences for both relativity theory and quantum physics. Given these wide-ranging ramifications, we have reason to ask if there is any part of the modern scientific worldview that would not be affected if a new theory of evolution became more widely accepted.

It's hard to know whether this holistic theory of a purposeful universe will gain ground in the professional scientific community. When scientists with Christian backgrounds some fifteen years ago launched the idea of intelligent design in biology,[23] this was met with a massive and uncompromising rejection, even though the idea probably seemed fairly obvious to most people. With this in mind, and considering the vast consequences it would have for science in general to accept the Tree of Life theory, it is anybody's guess how it will be received. It is in fact doubtful that the professional scientific community under any circumstances would be willing to recognize a theory that provides an alternative to Darwinism, since this provides its ultimate foundation. Possibly too much prestige has already been invested in Darwinism to allow for a fundamental revision. Nonetheless, the theory presented in this book is different compared to previous challenges to Darwinism in that it does not content itself with criticizing it, but in fact provides a new theory that integrates the evolution of the universe in a much wider context than has previously been proposed. This has required breaking some significant taboos of current science, however, notably the one against presenting natural processes as purposeful and the one against presenting theories that imply that a Creator might exist. Hence the response in the professional scientific community to the Tree of Life theory is unpredictable and may not necessarily have very much to do with its actual content.

Asking whether scientific theories really are as powerful in shaping the course of a civilization as is sometimes claimed is certainly warranted. Even a theory like Darwinism, which has commanded the unflinching support of almost all the world's scientists, has in some significant regards not been embraced. We may note for instance that some 90 percent of the world's population still believe in some higher spiritual reality, despite the denial of this by leading scientists. What is more, those who may professionally staunchly defend and promote Darwinism are not likely to live in accordance with that theory. When they come home to their loved ones, they almost certainly do not treat them as physicochemical whims of nature, but as human beings. Human feelings are mostly stronger than scientific theories, it would seem.

Having said this, I still think it is true to say that Darwinism has had very significant consequences for life, and at the very least for the course of science during the past 150 years on this planet. It is undeniable that this theory has influenced people when those who supposedly hold the most knowledge in modern societies—lauded scientists—have used their authority to consistently reject the intuitions of people at large, who may not have the time or inclination to dig deeply into the nature of biological evolution. It seems however that for people to have been willing to go along with Darwinism, there must have been some existential or emotional gain. Possibly this may have been a sense of supremacy on the part of the individual, who, if the inherent intelligence of the universe is denied, can instead celebrate his or her own intelligence. More broadly, among people at large, I feel that the denial of a spiritual origin to humankind has allowed many to enjoy a sense of freedom of responsibility from the consequences of their actions. After all, if we are only whims of nature, to whom, or for the manifestation of what higher purpose, are we accountable? To point to natural selection as a creative force in the universe, as many Darwinists in fact do, is like having a hidden agenda to create a religion, one that does not include a relationship to the source. The question now is whether our overpopulated planet can afford to continue with such a lack of accountability. Or, more specifically, will taxpayers all over the world continue to pay for biological research that is designed to hide not only their true origin, but also their purpose? Why should a particular form of atheist religion have the exclusively privilege to dominate the modern educational system? The assertion, commonly made by biologists and other scientists, that Darwinism is not afflicted with any problems breathes the same blocking of critical thinking that four hundred years ago characterized the Catholic Church. The price we have paid for the illusory freedom offered by Darwinism is the loss of a realistically based sense of higher purpose and of the sacredness of life.

I feel therefore that it is important to present a theory that can be of value to people in general as an expression not only of a new biology, but in fact of a new science. I want to offer it as a scientifically based understanding of evolution that does not contradict what most people intuitively know is right, namely that life is not an accident. In so doing I hope to rekindle an appreciation for the beauty of universal design and to reactivate the fascination that science provides. My sense is that many people feel a great interest in, even a thirst for, meaningful science that asks the big questions about life and yet remains faithful to the Cartesian ideal of fact and logic.

At the same time, because of the nature of the Galactic Underworld in which we currently reside, people have a broad, common desire to experience a fundamental oneness of humanity, and many are looking for science to provide an understanding of the unity that pervades all of creation. While the Tree of Life theory was not developed with the intention of responding to such a desire for oneness—it was prompted by the need to find out the truth about biological evolution—it seems that an understanding of the *common origin* of mankind is now truly at hand. The consequences of these discoveries, which mean that we literally have a representative of the Tree of Life in each one of our cells and that each and every one of us is created in the image of the Tree of Life, are far-reaching and affect all aspects of our existence. As we see that the Tree of Life is predominant in creation, it becomes understandable that our existence is part of a meaningful, unified whole. What is more, we may realize that because each of our cells is equipped with a microformatted Mayan calendar, humanity shares, and has shared throughout its existence, a common destiny and purpose. Regardless of race, religion, gender, political or sexual preference, or anything else, we all share an origin in the Tree of Life and a domain beyond our material existence. We are, in other words, something more than only the body and mind that are immediately apparent, and in our present existence we have through this common origin been given the possibility of experiencing beauty, compassion, and intelligence.

For those who accept the essentials of the new theory of universal evolution, and in particular biological evolution, this is bound to have extensive consequences for how they relate to the world and the many dichotomies that we have become used to as the chief philosophical reference points of our lives: atheism/theism, materialism/idealism, vitalism/functionalism, creationism/evolutionism, complementary/ mainstream medicine, and scientism/spiritualism, to name just a few. In many cases such dichotomies are likely to be transcended, but probably even more commonly they will lose their former meanings. For example, conservative religious groups in the past have typically defined themselves as antievolutionist and have rejected the notion of evolution. It now appears however that evolution really is a manifestation of creation and that the universe and its species have been intelligently designed with the purpose of evolving. In this new perspective, someone who is an evolutionist will by necessity at the same time also be a creationist, and vice versa. Thus I think it is accurate to say that the unification of evolutionism and creationism becomes a logical consequence of this new theory of biological evolution. I feel it is not an accident that the potential unification and the potential transcendence of old conflicts between science and religion are the result of the new theory of biological evolution. For all that we know, quantum physics and string theory in isolation have little or nothing to tell us about our origin or purpose unless they are integrated in a larger context with the evolution of life. It is only an understanding of biology that may truly explain who we are and define our place in the universe.

What has been revealed here is not just the basis for the integration of all science, but more fundamentally the basis for the insight that we all share the same origin in a purposeful universe. The fact that this theory is not linked to any particular religion and doesn't advocate any particular spiritual practice, but is more or less consistent with them all, means that it may help create a path to Oneness, a blending of many different traditions. Not as a nice thought, but a reality. It seems also that from the standpoint of the present theory, a strong case can be made that all aspects of the universe have one single source, and so *fragmentation either in scientific disciplines or in spiritual traditions no longer seems like a meaningful path toward understanding reality.* The entire observable universe comes out of one creation, brought about, or mediated by, the Cosmic Tree of Life, creating vibrations that all of us resonate with, and it is only logical that all of our science, if all of its branches are to be meaningfully connected, needs to be integrated into the study of this one reality.

It has here, I believe, been proven beyond a doubt that the universe and all of its biological organisms, including human beings, have been created by cosmic design, mediated, or executed, not only by the Cosmic Tree of Life, but also in the image of the Cosmic Tree of Life. Of course, this raises the question of the relationship between this creative intelligence and God. Is the Cosmic Tree of Life the same as God, which many ancient worshipers may have believed, or is there a transcendent God behind the execution of intelligent creation by the Cosmic Tree of Life? Is there, in other words, a God who is plucking the strings of the Cosmic Tree of Life to create a universal symphony? There are several possible answers to these questions, and while our findings are certainly consistent with the existence of a transcendent God, I do not feel that we have provided any empirical evidence of God's existence. It is in fact very doubtful if this would ever be possible. Yet, while it is possible that there is no God behind the Cosmic Tree of Life, and that creativity is simply the inherent nature of the Cosmic Tree of Life and the universe that we are living in, there is nothing in the present theory that speaks against the existence of a Creator God. I feel at this point we have come to the limit of the domain that may be scientifically investigated, and even if the Tree of Life theory has pushed our knowledge considerably into what previously would have been considered a metaphysical domain, the ultimate mystery about existence still remains. Arguably this may always stay outside of the limits of what we may intellectually comprehend. Nonetheless, I think that the criteria have been met for a complete theory, which was presaged by Stephen Hawking, who said, "All, philosophers, scientists, and just ordinary people, will be able to take part in the discussion of the question of why it is that we and the universe exist."[24]

Thus even if several different interpretations could be made regarding the relationship between the Cosmic Tree of Life and God, if any, I would like to say that my own is that Divine Intelligence manifests as the Cosmic Tree of Life for the purpose of creating a universe. Such an interpretation would mean that we are created not only in the image of the Cosmic Tree of Life, but also in the image of God, and I believe that what furthers the destiny of humankind is to see this image in others and to develop our consciousness about this and from this.

Consciousness is a necessary part of every interpretation of quantum theory. The reason for this is that quantum theory really is an aspect of the Tree of Life theory and that the Tree of Life exists in quantum states, which, as we have seen, correspond to states of consciousness. The origin of consciousness is however perhaps best understood from its etymology; *con* means "with" and *sciere* means "to know," so *consciousness* means "the state of knowing with." Thus in order to understand what consciousness is, we must answer this question: Knowing with what? The Swedish word for consciousness, *medvetande*, may provide a clue to answer this question, since in this language the word for knowing, *veta*, is related to *ved*, the English *wood*, meaning that in this translation consciousness is about "knowing with wood," and thus really knowing with the Tree of Life. It then takes two, the human being and the Tree of Life, for human consciousness to arise. Consciousness is a relationship rather than an attribute of a living entity and can only emerge in such a relationship. If the future of humanity is in the evolution of consciousness, it must then also be in the love for the Tree of Life and the Divine Intelligence that has generated this. Thus even if there seems to be no evidence that the Tree of Life ever intervenes to alter the course of evolution, it is fully possible that we—through enhancing the quality of our resonance with it through prayer, meditation, or otherwise—may alter our own perception of reality and so influence the manifestation of the cosmic plan.

Given the existence of the Tree of Life as the central creative principle of the universe, there is no reason to be surprised at the fact that humankind throughout the course of its existence has developed spiritual practices and religious traditions to answer the question of how we are to relate to this creative source. These may range from Siberian shamans climbing trees in a trance state to travel to the otherworld, to scripture-based creation stories in religions that define ethical rules and the practice of rituals. Certainly it is natural to ask what this creative force that dwarfs us in all respects wants us to do, and there is no reason at all to wonder why religions exist to begin with, whether we choose to adhere to them or not.

Let us now look at one more aspect of our origin.

THE RETURN OF THE SOUL

Now, see how the image of man and woman stems from the divine proportion. In my opinion, the propagation of plants, and the progenitive acts of animals are in the same ratio as the divine proportion, or proportion represented by line segments.

JOHANNES KEPLER

With few exceptions the existence of a soul is something that most religions and spiritual traditions in the past few thousand years have taken for granted. Eastern religions, such as Buddhism and Hinduism, often have a view of a soul that reincarnates in different bodies, and the independence of the soul from the body is also recognized in shamanistic practices such as soul retrieval. In the Western world the belief in the existence of a soul can be traced to Greek philosophers, while the Jewish religion, which placed little emphasis on the afterlife, traditionally tended to look upon the soul and body as inextricably linked. It is said that it was the apostle Paul, who, because of his many contacts with Greek philosophy, introduced the concept of a soul into the Christian religion. Muslim philosophers also seem to have been directly influenced by the Greeks in this regard. Many people on this planet, belonging to the major religions or otherwise, have thus had a view of their own spiritual essence, their soul, as something that has a tangible existence distinct from their body and has a special link to the spiritual domain.

Generally, the phenomenon of a soul has only rarely been dealt with by modern science, and it could in fact not easily be discussed because of its current view of the origin of the human being. While current biology recognizes that the sperm and the egg from which a human being is conceived have very complex internal architectures, it is probably fair to say that in principle, biologists regard the two as bags of biomolecules. Such a view of the origin of individual human life really precludes the notion that we might be endowed with souls, since it is not easy to see where the spiritual dimension would enter into the bags.

Likewise, the capacity for expressions of compassion or love would be unlikely to emerge merely from what would be generated by a mixture of two bags of biomolecules, even though it is sometimes said that our emotions and feelings of love can be reduced to brain chemistry. From the perspective of this new theory of biological evolution, it seems however that compassion and love may have a different origin. To explore this, we may first notice that human beings generally feel the most compassion toward other human beings, or possibly their pets or other fellow mammals, which also happen to be the beings that most clearly have been created in the image of the Tree of Life. Yet to varying degrees we also feel compassion toward lower animals, and some of us may be hesitant even to kill spiders. As we go to plants, fungi, and bacteria it seems however that we pass the limit of what our compassion may be extended to. One would be hard pressed to find a person who has such a strong compassion for plants that he or she for ethical reasons would stop eating vegetables. It may be that the basis for our varying degrees of compassion toward different species is whether their cells have centrioles, since as I pointed out earlier, plants, fungi, and bacteria, toward which we have only very limited compassion, lack centrioles. What is implied here is that it is the degree of expression of the

Tree of Life in an organism that evokes compassion in us and does so increasingly the more clearly the organism has been created in its image. Hence, I think the perceived presence of the Tree of Life is ultimately what provides the basis for human ethics. Even though the ideals of ethical conduct will vary considerably in different cultures, I feel that in the recognition of the Tree of Life in other beings, there is a basis for ethics at the very root of our existence that may not previously have been recognized. I think it is important to note that to many people, the very fact that love and compassion exist at all in the universe serves as evidence of the existence of God.

In this context, we may also remember that in Genesis, the Tree of Life is referred to as the Tree of Knowledge of Good and Evil, and maybe we have by this observation discovered the root of the name. On some level we may all have a sense not only of the beauty, but also of the sacredness of the Tree of Life to the extent that it evokes compassion for organisms that are created in its image. I believe that resonance with the Tree of Life brings an enhanced life to an organism and provides it with a portal to a sacred spiritual realm, which is the basis for compassion. Even if human beings may harm, violate, or even kill each other, this will always hurt ourselves, since such acts harm the image in which we ourselves have been created. Although some may deny the guilt associated with violating another, such denial will by necessity come at the price of the quality of the spiritual contact with the source, and it is on this realization that norms for ethical and compassionate conduct may have been developed.

This also has implications for the current time. As we approach the highest quantum state of the Cosmic Tree of Life on October 28, 2011, and the time after this, when the cycles humans currently perceive come to an end, I believe that humanity's enhanced awareness will contribute to an increased compassion. This will be much needed, as the transition from a growth economy to a sustainable economy necessary for our survival is likely to cause considerable perceived or real hardships for billions of people expecting their material lot to improve.

Love and compassion are usually seen as emanations of the human soul, and we may then wonder if the existence of a soul may be accommodated by this new scientific theory of evolution. To explore this we may look at how the individual human life begins and whether this could also include the birth of a soul. A few hundred years ago this would have been taken for granted, and a typical view was that the sperm contained a homunculus (figure 9.6), a minuscule soul and body of a potential human being that after it had entered the egg would develop into the fetus and then later into the child and the adult. In those early days of cell biology, some investigators even claimed to be

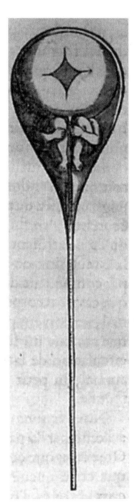

Figure 9.6. Homunculus. Seventeenth-century drawing of a homunculus, which was seen as a little human being carried by the sperm. Note the four-directional star in the sperm head, possibly indicative of the idea that something was in there providing directions for the development of the homunculus. From Nicolas Hartsoeker, Essay de dioptrique *(Paris: 1694), with permission from College of Physicians in Philadelphia.*

able to see such a homunculus in the microscope, but later these notions came to be regarded as prescientific superstitions. But as so often happens in the history of ideas, things come back at a later point, albeit in a new form. Thus while the homunculus concept may not have been literally true, it is as we shall see more true than the modern scientific idea that a new life emerges by the mixing of two bags of biomolecules—one from the sperm and one from the egg.

As we all know, neither the sperm nor the egg has by itself the ability to develop into a human being. It takes the unification of the two for the conception of a new life. I have however previously mentioned that the centriole plays such a significant role in development that its implantation in frogs may be used for conceiving virgin births. We may then naturally ask how the centriole emerges in the fertilization process in humans. This is relevant to ask, since neither the egg nor the sperm holds a centriole, and in fact the two of them need to be unified in order for a centriole to emerge and development to begin. In mammals the centriole is not present in the sperm cells, which unlike the egg cell, are mobile and for this employ their flagella to generate their movement. Interestingly, however, the anchoring point of the flagella is provided by a basal body, which is what serves as an engine for the tail propeller. Such a basal body is however interconvertible with a centriole and is thus mostly regarded as the same organelle. When a sperm head enters the egg, the very same basal body that had been driving its tail is converted into a centriole,[25] which then starts to organize the fertilized egg (see figure 9.7). Without a centriole it would not be possible for a fertilized cell to connect with and receive the Halographic information from the Cosmic Tree of Life that is necessary for a multicellular organism to develop. In my own view, the emergence of the centriole in the fertilized egg, and not the unification of the genes, is the critical event bringing about fertilization and is what commits the fertilized egg to the development of an embryo and a fetus.

This emphasis on the role of the centriole in fertilization is not new, even though it may have been forgotten. Theodor Boveri, who

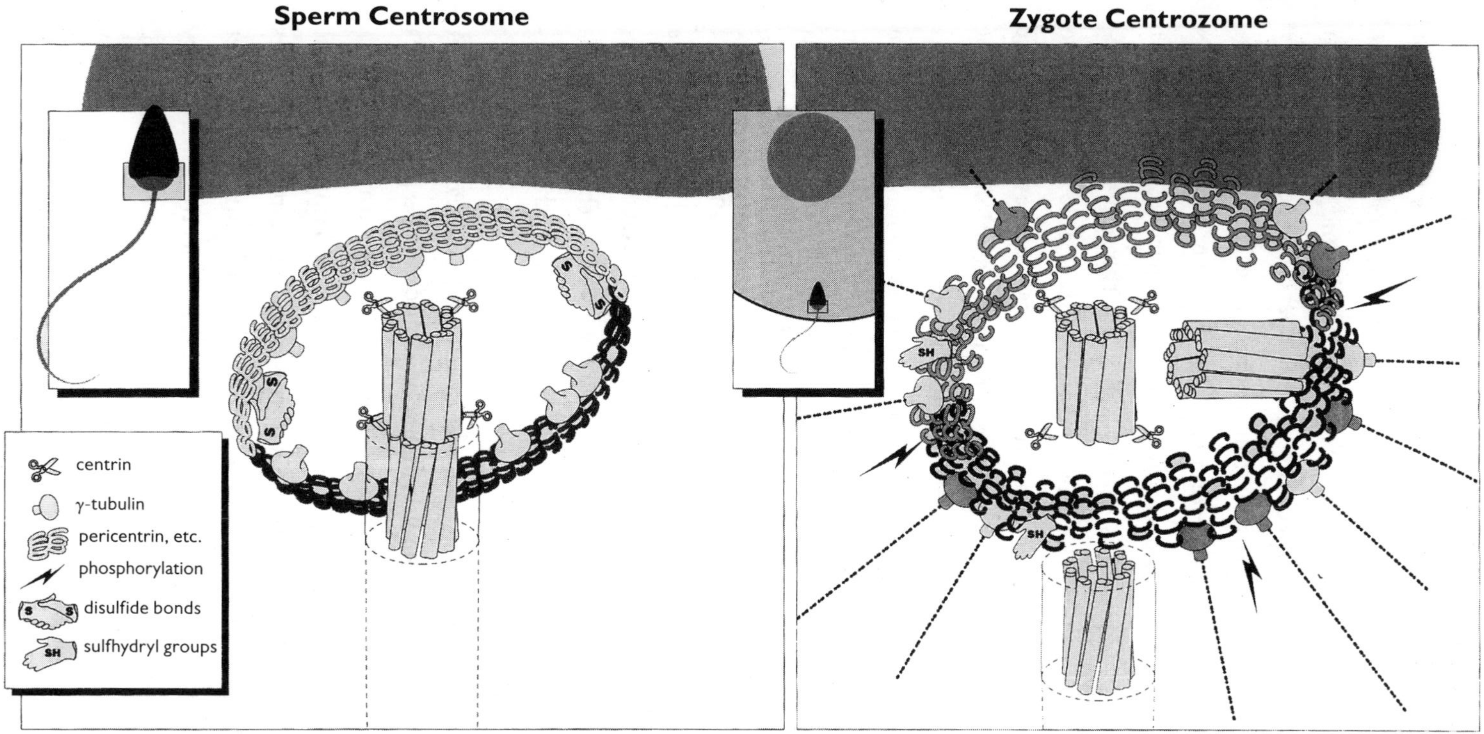

Sperm Centrosome

Zygote Centrozome

centrin
γ-tubulin
pericentrin, etc.
phosphorylation
disulfide bonds
sulfhydryl groups

Figure 9.7. The basal body of the sperm (left) is transformed to a centriole (right) at fertilization. As the sperm enters the egg and fertilization takes place, the basal body that has been serving as an engine for the movement of the sperm tail (see the rectangle in the insert in the left picture) is transformed into a centriole with a perpendicular crosslike structure, shown in the picture to the right. From G. Schatten, "The Centrosome and Its Mode of Inheritance: The Reduction of the Centrosome During Gametogenesis and Its Restoration During Fertilization," Dev. Biol. 165 (1994): 299–335. With permission from Elsevier.

discovered the centriole, in 1887 developed an amazing theory about fertilization: "The ripe egg possesses all of the elements necessary for development save an active division center. The sperm on the other hand possesses such a center but lacks the protoplasmic substratum in which to operate. In this respect the egg and sperm are complementary structures; their union in syngany thus restores to each the missing element necessary to further development."[26] What may be added here is that the reason this active division center, the centriole, plays such a critical role in development is that it establishes the necessary connection with the Cosmic Tree of Life and from this is able to access the wave form information that regulates the process of development. It is through this resonance that a blueprint for the relational pattern of the centrioles in the different cells can be downloaded from the world beyond. This organismic wave form that defines the relations among the centrioles in the different cells that define the body plan of an organism and how its embryological development will take place and result in a mature organism. Thus even if the sperm does not bring a homunculus, it does indeed bring the resonance unit, the centriole in whose image the body plan of the growing individual is developed. In this way we may understand that it is the establishment of a contact with the creative center of the universe that ultimately gives rise to the growth of a fetus with a soul and an infant that later is born with the Milky Way still in its eyes.

This view of how fertilization takes place gives us reason to revisit the quote from Johannes Kepler that introduced this section. Kepler's idea that creation reflected God's geometry has sometimes been ridiculed by modern scientists, but knowing what we know now about centriole biology, we have every reason to stand in awe not only of his intelligence, but even more so of his intuition. Four hundred years after he wrote those words, only after gaining access to electron microscopy and liberating ourselves from the confusing randomism of Darwinism, can we verify that Kepler was exactly right. Kepler somehow simply knew that "the progenitive acts of animals are in the . . . divine proportion . . . represented by line segments," or in other words, that something like the centriole provided line segments in accordance with the divine proportion, from which stem the images of man and woman. What Kepler showed is how a great scientist, who for some reason has a much-developed sense of resonance with the Cosmic Tree of Life, will simply know things long before the empirical evidence is present. This is not to say that good science can ever be performed without verification through observations and experiments, but it does tell us that the source of knowledge is always accessible through resonance with the Cosmic Tree of Life. If we ever wanted proof that humans can simply know the way things are without actually making measurements, this quote from Kepler is hard to beat.

If the biological manifestation of a wave form requires the antenna of a centriole for reception, we may wonder how a wave form is expressed when such an antenna is not at hand. The natural answer is that it remains in a form that human beings over the millennia have been referring to as part of the spirit world, particularly in shamanic traditions. If we accept the quantum nature of the Cosmic Tree of Life, it becomes understandable that this might also generate many wave forms that humans have experienced as angels, naguals, ghosts, or otherwise. These would then simply be the expressions of wave forms that do not get anchored in biological matter through the resonance of the centriole. Clearly the human experience of such nonmanifest wave forms are influenced by the particular frame of consciousness they are living in. At the current time, in the Galactic Underworld, orbs and other light phenomena seem to be a prevalent wave form of this nature, visible in many photographs. This would in principle not be different from how in earlier Underworlds the teleportation of quantum states from the Cosmic Tree of Life gave rise to Halo phenomena on a galactic and geophysical level. The only difference would be that orbs create a more subtle effect at our current level of evolution. To discuss these matters in depth would require another book. Suffice it to say however that the current theory is fully compatible with the existence of wave forms that never manifest biologically and can be referred to as spirits or nonphysical light phenomena.

In the view that emerges here, the origin of the individual human life is primarily to be found in the wave form, based on the geometry of the Cosmic Tree of Life, and only secondarily in the biomolecules inherited from parents. It is from the Halographic projection onto the fertilized egg from the Cosmic Tree of Life that a living soul that is whole and unique comes into being. Hence, not only are human individuals whole, they are also unique creations by the Cosmic Tree of Life. The soul primarily has its origin in a wave form emanating from the Cosmic Tree of Life, in whose image it has been created. The soul of the child thus does not come from the parents, and they do not own the being newly arrived to the world. Instead it is for them to nurture and assist the development of this soul that came to them from a higher source. Such a view means a fundamental reframing of the origin of our lives as individuals. This is why we must look upon each newborn child as a unique soul with a unique destiny. This is why we must fully appreciate that children have ideas of their own and destinies that may be quite different from those of their parents or even what they may be able to conceive. If human beings did not each individually have a unique origin in the Cosmic Tree of Life, we would be little more than Pavlovian dogs, conditioned by the bags of biomolecules that we had inherited from our parents and by the events in our early lives.

As we know, children are not like Pavlovian dogs, but sometimes turn out very differently even when they are reared under very similar circumstances. Although most good parents care about the individuality of their children, regardless of any biological theories, I feel it should be pointed out that with the new theory of biology presented here, the higher origin, uniqueness, and autonomy of the soul of each child is not just some nice thought, but simply the truth. While in the current paradigm psychological science considers only two general frames of influence, genes and environment, for determining how the life of a child will turn out, we can now see that this perspective is incomplete, since it ignores the existence of the spiritual essence. Thus while environment and genes certainly may modify the life path in many different ways, neither of them explains why a person has any intention of moving forward in life to begin with or any ability to create something new. The reason that we all to some degree experience life as a kind of project developing over time is that evolution, also of an individual life, is an inherent aspect of this creation, which is driven forward by the invisible shifting of the yin/yang polarities of our wave forms. This in a broad sense explains how our individual lives gain their direction. Even though we may sometimes change the path of our life, the only reason we walk a path to begin with is that we are under the influence of such shifting Halos. The origin of our lives in the higher realm of the Cosmic Tree of Life is the reason that so many children, in spite of sometimes very difficult conditions in childhood, show that they are not the mere results of genes and environment, but also of something more that leads them to have productive lives. At a deeper level most of us know that we have an origin in a sacred realm beyond material reality, and this is what our souls help us remember.

Consistent with this origin of the human soul in the Cosmic Tree of Life there are, as was mentioned in the first chapter, several ancient creation stories where it is held that the first human being was born from a tree. With the identification of the centriole as the Cellular Tree of Life, and the recognition of its emergence as the key fertilization event, I feel these ancient views have been shown to be literally true. Two pictures shown here (figures 9.8 and 9.9a) are both from ancient Mexico and illustrate the role of the Tree of Life in the life cycle conceptualized by these cultures. Figure 9.8 shows the famous sarcophagus lid on the tomb of Pacal in the Temple of the Inscriptions in Palenque in Chiapas, Mexico. This lid shows how Pacal at his death falls into the roots of the World Tree, which is apparent behind him as a cross that covers most of the lid. This relief illustrates the Mayan notion that as you die you return to the World Tree (or actually to the quadripartite god at its roots). In the second picture (figure 9.9a), from an ancient Mixtec codex, we can see not

Figure 9.8. The World Tree in Pacal's sarcophagus lid. The sarcophagus lid of Pacal's tomb in the Temple of the Inscriptions in Palenque shows how Pacal, who was a significant member of the ruling dynasty in Palenque in the seventh century, falls into the roots (the quadripartite god) in death. The World Tree may be seen as a cross behind him. Courtesy of Linda Schele and David Freidel.

only the head of a deceased person sinking into its roots, but at the same time also a man being born from the crown of the World Tree.

Whether in the second picture the two are actually one person reincarnating or if this illustrates a more general recycling by the Tree of Life from a common pool of souls we do not know. Regardless, the pictures demonstrate the ancient Mesoamerican view that the World Tree was the portal between the world of the dead and that of the living. Our souls were not only born from it, but also return to it in death. Also in Viking mythology the first man, Ask (meaning "ash" in Swedish), was given his name by the huge World Tree Yggdrasil, from which he had been born. Thus in two contemporary cultures as diverse as the Mayan and the Viking, the origin of humans is recognized to be the Tree of Life. It seems possible that the Christian notion of Jesus dying on the cross and then resurrecting could be an elaboration of the same relationship to the Tree of Life, since it does provide the portal to the otherworld and a possible eternal life of the soul there. This interpretation seems especially likely considering that according to legend, the Golgotha cross was made from the wood of the Tree of Life in the Garden of Eden.

We should also notice that the Tree of Life in figure 9.9a, from which the human entered the world, looks similar to the Christian Tau Cross. (As a curiosity we may here note that in the professional mathematical literature, the golden ratio is symbolized by the Greek letter τ [tau] rather than ϕ [phi]). The Tau Cross is sometimes also called an Old Testament Cross and was used especially by St. Francis. The Tau Cross also appears in the painting in figure 9.9d as the source of Jesus being born into the Holy Family, which may be seen as symbolic of how the centriole mediates the soul of the newborn child. The Mayan World Tree symbol in the day-sign Ik* is essentially identical to the Tau Cross, and so the two symbols share the basic structure of the centriole and

---

*The Ik symbol among the Mesoamerican represents the wind god, who among the Aztec is called Ehecatl, the wind aspect of Quetzalcoatl, the bringer of civilization. Wind here does not refer to the shifting of air in the physical sense, but to movements emanating from the World Tree.

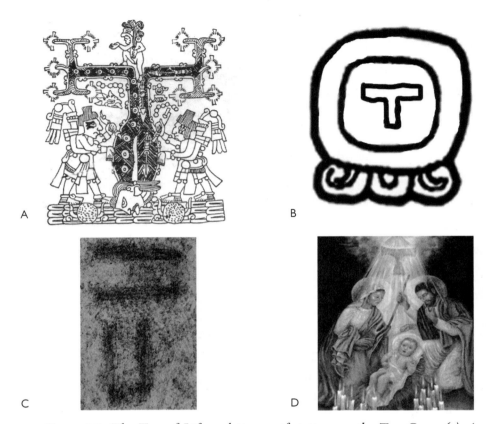

Figure 9.9. The Tree of Life and its manifestations as the Tau Cross. (a) A man is born from the crown of the Tree of Life with (another?) man entering its roots at death, from the Mixtec Codex Vindobonensis. It is unclear if this means that in the ancient Mesoamerican view the individual soul was seen as reincarnating or if the Tree of Life was seen as a portal to the Otherworld, to which souls returned at death. (b) The glyph for the day-sign Ik in the Sacred Calendar, where it is a symbol of the wind god and Quetzalcoatl. This day-sign symbolizes the spiritual winds that emanated from the Tree of Life. (c) Photograph of a centriole, the Cellular Tree of Life. (d) The birth of Jesus in the Holy Family. Painting from Santo Pedro Hermano, Antigua, Guatemala.

the Tree of Life. The Tree of Life, manifested in the fertilized egg as the centriole, provides the point of entry of the soul. The Tree of Life may also be seen as a point of return as the body dies, which symbolically is represented in different religions for instance by the Christian cross at burial sites. The belief in a cycle of birth, life, death, and rebirth as well as in an eternal soul, presented with some variations in the world's religions, is not inconsistent with the new theory of biological evolution.

What we have arrived at here is thus a biology with a soul, a scientific understanding of biology that allows for the existence of what most of us would consider to be our very essence, something that has an origin in another realm and is not inextricably linked to our physical bodies. Different cultures have had different ideas concerning how the soul is related to the body, and it is not my intention to present any idea as right or to prescribe how to look at the relationship between body and soul or the fate of the soul after death. Empirical evidence is hard to find on these matters—meaning that many different interpretations are possible—and so we are again facing a mystery where, at least at the present time, science is of little avail.

In ending, it is natural to ask what the purpose of this birth, death, and rebirth of souls may be from the perspective of the universe and the Cosmic Tree of Life. We have shown that the universe is purposeful, but we have not discussed what its purpose is. Maybe the answer lies in an Islamic hadith wherein Allah says, "I was a hidden treasure and wanted to be known."* This may be the very reason that the universal process of evolution is leading to a point, October 28, 2011, when the climb up the Cosmic Pyramid will be complete, and a consciousness of light will be attained. This date is then not simply a date marking a point in the linear flow of time, but a reflection of the quantum state 13.13.13.13.13.13.13.13.13. 13 Ahau, which is the highest possible quantum state of the universe. All that the Mayan calendar coming to an end truly means is that the highest quantum state of the universe will be attained. Only by the establishment of a final stable quantum state can the prospect of a millennium of peace be realistic; and, if this

---

*This is the beginning of a *hadith qudsi*, or sacred hadith, which are words of God outside the Qur'an. This supposedly was the divine response to the prophet David's query about the purpose of creation.

is true, then all models suggesting that the Mayan calendar is based on astronomical cycles would be disqualified. At that point in time, when all the nine cosmic forces associated with the Nine Underworlds are manifesting fully, we will be ready for something entirely new. We are then meant to reflect a quantum state that would allow us to perceive the unity of all things and to attain a new level of resonance with the divine source. If this is to be experienced on our particular planet, it calls for a very rapid transformation of our life and an understanding of where we come from and where we are meant to go. If we are to return to the Garden of Eden at a new and higher level, as I believe is planned, we need to be aware of the Tree of Life at its center and relate to this as the origin of the living universe.

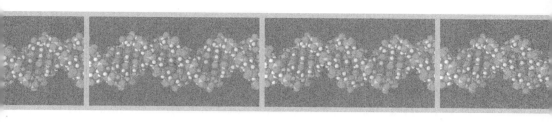

Glossary

Ahau: twentieth day sign of the uinal, meaning "light/lord"

ahauob: Mayan rulers (lords)

alautun: period of 63.1 million years

amino acids: components of proteins

anthropoid: a suborder (Anthropoidea) of higher primates, such as apes.

anthropology: study of the development of human beings

Aztec: name commonly used for the Mexicas, based on their origin in Aztlan

baktun: period of 394 years

big bang: beginning of the creation of the universe; a giant "explosion" in which matter was first formed from light some 15 billion years ago

biosphere: region of the earth's crust and atmosphere harboring living organisms

celestial: pertaining to the heavens

Cellular Creation Cycle/Cellular Underworld: first level of creation of thirteen hablatuns, beginning 16.4 billion years ago

center of the world: Yaxkin, the Mayan center of the Four Directions

centriole:

Chalchiuhtlicue: Aztec goddess of water, ruler of the third day of the trecena

channeling: receiving information from a higher spiritual source

Chichén Itzá: major Mayan temple site on the northern Yucatán peninsula

Cinteotl: central Mexican god of maize (corn) and sustenance, ruler of the seventh day of the trecena

classical Maya: Maya who lived in classical times, about 200–850 CE

codex: ancient Mesoamerican book, usually made from bark

Codex, Dresden: postclassical Mayan codex used by the German librarian Förstemann to elucidate the calendrical system of the Maya

consciousness: invisible boundary that separates inner and outer reality

Cortés, Hernan: Spanish conqueror of Mexico

cosmic consciousness: all-encompassing consciousness

cosmic pyramid: nine-story pyramid

crust: surface layer of the earth

daykeepers: special Mayan persons responsible for keeping track of the passage of days

dark matter: nonluminous matter not yet directly detected by astronomers that is hypothesized to exist to account for various observed graviational effects.

day-signs: the names of the deities ruling the days of the uinal

divine plan: the exact progression of spiritual energies determining the evolution of consciousness

endosymbiosis: symbiosis in which a symbiont dwells within the body of its symbiotic partner

epigenetics: relating to, being, or involving changes in gene function that do not involve changes in DNA sequence

equinox: one of two days in the year when the day and the night are equally long

ethereal: nonmaterial

exobiology: a branch of biology concerned with the search for life outside the earth and with the effects of extraterrestrial environments on living organisms.

extrasolar planets: planets around other stars than our own sun

extremophile: an organism that lives under extreme environmental conditions

Familial Creation Cycle/Familial Underworld: third level of creation, developed by thirteen kinchiltuns beginning 41 million years ago

Feathered Serpent: main Mesoamerican deity, symbolizing light and the creation of culture and civilization; also called Quetzalcoatl or Kukulcan, the Plumed Serpent

First Father: Mayan deity that raised the World Tree; also known as the maize god

Galactic Creation Cycle/Galactic Underworld: eighth level of creation, developed by thirteen tuns beginning January 5, 1999

galaxy: star system; sometimes, as in the case of the Milky Way, spiral shaped and with hundreds of billions of stars

global brain: organization of the planet according to the functionalities of the human brain

glyph: symbol of Mayan writing

hablatun: period of 1.26 billion years

hemisphere: half sphere

heterogeneity: consisting of dissimilar or diverse ingredients or constituents: mixed

hierarchical: arranged or organized in a graded or ranked series of different levels

holographic: relating to a structure in which the whole is microscopically reflected in each of its parts

holographic resonance: relay system for synchronistic information transmission from the macrocosmos to the microcosmos

homogeneity: of uniform structure or composition throughout

Homo habilis: first species of the family *Homo,* living about 2 million years ago in East Africa

inner core: region of the earth's center

Ix: fourteenth day-sign of the uinal, meaning "jaguar"

kalabtun: period of 160,000 years

katun: period of 19.7 years (7,200 days)

kin: sun, period of one day

kinchiltun: period of 3.2 million years

Long Count: The Long Count calendar identifies a date by counting the number of days from August 11, 3114 CE in the proleptic Gregorian calendar or September 6 in the Julian calendar (−3113 astronomical dating)

macrocosmos: large-scale manifestations of the cosmos

mammalian brain: the lateralized brain structure of mammals; its functions differ between the two hemispheres

Mammalian Creation Cycle/Mammalian Underworld: second level of creation, developed by thirteen alautuns beginning 820 million years ago

mantle: region between the earth's crust and its core

Medicine Wheel: wheel that includes the Four Directions, used for ceremonial purposes by North American Indians

Mesoamerica: "Middle America," archaeological term for the cultural area ranging from northern Mexico to Honduras

metaphor: verbal image used to illustrate an abstract phenomenon

microtubules: any of the minute tubules in eukaryotic cytoplasm that are composed of the protein tubulin and form an important component of the cytoskeleton, mitotic spindle, cilia, and flagella.

nagual: sometimes a totem; a symbolic manifestation of the Otherworld

National Creation Cycle/National Underworld: sixth level of creation, developed by thirteen baktuns with a total duration of 5,125 years

New Age: modern movement prophesying the advent of a New Age

New Jerusalem: new world emerging at the end of time as described in the Book of Revelation

number 7: holy number in Mesoamerica, the number of Light pulses in each Underworld

number 9: holy number in Mesoamerica, the number of Underworlds

number 13: holy number in Mesoamerica, the number of Heavens

number 108: holy number in the Hindu and Buddhist traditions; Indian sages often take this as one of their names

October 28, 2011: end date of the Divine Creation cycles

octupole: Two electric or magnetic quadrupoles having charge distributions of opposite signs and separated from each other by a small distance

Olmecs: ancient Mesoamerican people emerging around 1500 BCE who inhabited the gulf coast of Mexico

optimal: best under a given set of circumstances

oscillation: wave movement

outer core: region between the earth's inner core and its mantle

Pacal: king of Palenque in the seventh century CE

physical biology: discipline consisting of the interaction of physics with biology

physical time: notion of time as being based on the cyclical movement of material bodies

pictun: period of 7,900 years

Planetary Creation Cycle/Planetary Underworld: seventh level of creation, developed by thirteen katuns beginning in 1755 CE

planetary midline: vertical arm of the World Tree, corresponding to longitude 12° East on the earth

Pleiades: group of stars in Taurus, part of Gould's Belt

Plumed Serpent (or Feathered Serpent): called Kukulcan by the Yucatec Maya and Quetzalcoatl by the Aztec polarity: duality creating tension

Popol Vuh: "Book of Advice," text of ancient origin relating the Mayan creation mythology, sometimes called the Bible of the Maya

postclassical Maya: Maya living in the period 850–1250 CE

precession: circular movement of the earth's axis

quadrupole: a system composed of two dipoles of equal but oppositely directed moment

quantum mechanics: physics of wave/particle duality

Quetzalcoatl: Aztec (Nahuatl) name for the Plumed (or Feathered Serpent)

Quiché Maya: group of Maya living in present-day Guatemala

Qur'an (Koran): holy scripture in the Islamic religion, consecrated by the Prophet Muhammad in 632 CE

Regional Creation Cycle/Regional Underworld: fifth level of creation of thirteen pictuns, beginning 102,000 years ago

reincarnation: idea that the human soul passes from life to afterlife to life again

Revelation: Apocalypse, the end scenario of creation in the Bible

Sacred Calendar: name given to the 260-day tzolkin

shaman: person with the ability to contact and see into the Otherworld

Shiva: Hindu god of creation and destruction

solar year: period corresponding to one revolution of the earth around the sun spring and autumn equinoxes: days of spring and autumn, respectively, when the duration of the day equals that of the night

stele (or stela): stone slab erected to celebrate a ruler (plural: stelae)

subconscious awareness: an awareness that is unknown to the conscious mind

superstition: belief without foundation

synchronicity: statistically unlikely event that appears meaningful and meant to happen

Teotihuacán: major ceremonial pyramid complex and commercial center outside of today's Mexico City; it flourished from about the time of the birth of Christ to the early eighth century CE

Tezcatlipoca: Aztec god of Darkness, ruler of the tenth day of the trecena and the thirteenth and fifteenth days of the uinal; the nemesis of Quetzalcoatl

Tikal: major Mayan temple site in present-day Guatemala

Toltecs: Mesoamerican people whose capital was Tula in the state of Hidalgo

Tribal Creation Cycle/Tribal Underworld: fourth level of *n*; a round of tones 1–13 creation of thirteen kalabtuns beginning 41 million years ago

uaxaclahunkins: Mayan name for an eighteen-day cycle

uinal: twenty-day period; a round of the twenty Day Lords

Universal Creation Cycle/Universal Underworld: ninth level of creation of 260 days beginning March 8, 2011

World Mountain: center of the earth in many ancient traditions

World Tree: perpendicular organizing structure for the creation of the cosmos

Xiuhtecuhtli: Aztec god of fire and time, ruler of the ninth day-sign of the uinal and the first day of the trecena

Yggdrasil: Norse name for the World Tree, imagined as a huge ash

yin/yang: Chinese names for the Dark/Light, female/male polarity of the cosmos

yin/yang fields: projections of yin/yang dualities onto the earth's surface

Yohualticitl: Aztec goddess of birth, ruler of the eleventh day of the trecena

Yucatec: originating on the Yucatán peninsula

Zapotecs: ancient Mesoamerican people who discovered the tzolkin; they lived in the state of Oaxaca and built the ceremonial center of Monte Alban

zenith: high point, when the sun is directly above a certain location

Notes

PREFACE

1. Stephen Hawking, *A Brief History of Time* (New York: Bantam, 1988).
2. Carl Johan Calleman, *Solving the Greatest Mystery of Our Time: The Mayan Calendar* (London: Garev Publishing, 2001); Carl Johan Calleman, *The Mayan Calendar and the Transformation of Consciousness* (Rochester, Vt.: Bear & Co., 2004).

CHAPTER I.
THE UNIVERSE IS NOT HOMOGENEOUS—
AND IT NEVER WAS!

1. Harold J. Morowitz, *The Emergence of Everything: How the World Became Complex* (New York: Oxford University Press, 2004), 39–43.
2. Steven Weinberg, *The First Three Minutes: A Modern View of the Origin of the Universe* (New York: Basic Books, 1988), 154.
3. G. F. Smoot, C. L. Bennett, A. Kogut et al., "Structure in the COBE DMR First Year Maps," *Ap. J. Lett.* 396 (1992): L1.
4. C. L. Bennett, M. Halpern, G. Hinshaw et al., "First Year Wilkinson Microwave Anisotropy Probe (WMAP) Observations: Preliminary Maps and Basic Results," *Astroph. J Suppl. Ser.* 148 (2003): 1–27; e-print: astro-ph/0302207.
5. M. Tegmark, A. de Oliveira-Costa, and A. Hamilton, "High Resolution Foreground Cleaned CMB Map from WMAP," *Physical Review D* 68 (2003): 123523.

6. A. de Oliveira-Costa, M. Tegmark, M. Zaldarriaga, and A. Hamilton, "The Significance of the Largest Scale CMB Fluctuations in WMAP," *Phys. Rev. D* 69 (2004): 063516.

7. J. Magueijo and K. Land, "Examination of Evidence for a Preferred Axis in the Cosmic Radiation Anisotropy," *Phys Rev Lett* 95 (2005): 071301; Katharine Rosemary Land, "Exploring Anomalies in the Cosmic Microwave Background" (thesis, University of London, 2006).

8. K. Land and J. Magueijo, "The Axis of Evil," http://arxiv.org/abs/astro-ph/0502237 (accessed August 3, 2009).

9. Dominik Schwarz, Universitaet Bielefeld, personal communication.

10. A. Cho, "A Singular Conundrum: How Odd Is Our Universe?" *Science* 317 (2007): 1848–50.

11. Ibid.

12. D. Hutsemékers, R. Cabanac, H. Lamy, and D. Sluse, "Mapping Extreme-Scale Alignments of Quasar Polarization Vectors," *Astron. Astrophys.* 441 (2005): 915–30.

13. M. J. Longo, "Does the Universe Have a Handedness?" e-print: astro-ph/0703325.

14. Ibid.

15. "Public to Join Search for Cosmic 'Axis of Evil,'" *New Scientist,* July 11, 2007, www.newscientist.com/article/mg19526123.300-public-to-join-search-for-cosmic-axis-of-evil.html (accessed August 3, 2009); see also www.galaxyzoo.org regarding the public participation study.

16. K. Land, A. Slosar, C. Lintott et al., "Galaxy Zoo: The Large-Scale Spin Statistics of Spiral Galaxies in the Sloan Digital Sky Survey," *Monthly Notices of the Royal Astronomical Society* 388 (2008): 1686–92.

17. Michael Longo, University of Michigan, personal communication.

18. Carl Johan Calleman, *Solving the Greatest Mystery of Our Time: The Mayan Calendar* (London: Garev Publishing, 2001), 107–8; Carl Johan Calleman, *The Mayan Calendar and the Transformation of Consciousness* (Rochester, Vt.: Bear & Co., 2004), 103–4.

19. Michael Talbot, *The Holographic Universe* (San Francisco: Harper, 1992).

20. Gene Fernandez, "The Maya Creation Story," www.theosociety.org/pasadena/sunrise/50-00-1/am-fern.htm (accessed August 3, 2009).

21. G. J. Whitrow, quoted in W. L. Craig, "The Teleological Argument and the Anthropic Principle," www.leaderu.com/offices/billcraig/docs/teleo.html (accessed August 3, 2009).

22. Martin Rees, *Just Six Numbers: The Deep Forces That Shape the Universe* (New York: Basic Books, 1999).

23. S. Weinberg, *The First Three Minutes, A View of the Origin of the Universe* (New York: Basic Books, 1988). The quantum vacuum has recently been highlighted in Lynne McTaggart's well-researched book *The Field: The Quest for the Secret Force of the Universe* (New York: Harper Perennial, 2003).

24. Stephen Hawking, "Quantum Cosmology," in *The Nature of Space and Time*, by Stephen Hawking and Roger Penrose (Princeton, N.J.: Princeton University Press, 1996), 89–90.

25. Martin Rees, *Just Six Numbers: The Deep Forces That Shape the Universe* (New York: Basic Books, 2000), 54.

26. Ibid.

27. Lee Smolin, "How Were the Parameters of Nature Selected?" http://online.itp .ucsb.edu/online/bblunch/smolin/oh/01.html (accessed August 3, 2009).

28. James N. Gardner, *Biocosm: The New Scientific Theory of Evolution: Intelligent Life Is the Architect of the Universe* (Makawao, Hawaii: Inner Ocean, 2003), 38–48.

29. Carl Johan Calleman, *Solving the Greatest Mystery of Our Time: The Mayan Calendar* (New York: Garev Publishing 2001); C. Calleman, *The Mayan Calendar and the Transformation of Consciousness* (Rochester, Vt.: Bear & Co., 2004).

CHAPTER 2.
TIME AND THE CALENDARS OF THE MAYA

1. Prudence M. Rice, *Maya Political Science: Time, Astronomy, and the Cosmos* (Austin: University of Texas Press, 2004).

2. See for instance Gordon Brotherston, *Painted Books from Mexico* (London: British Museum Press, 1995), 109–17.

3. Linda Schele and David Freidel, *A Forest of Kings: The Untold Story of the Ancient Maya* (New York: Morrow, 1990).

4. David Freidel, Linda Schele, and Joy Parker, *Maya Cosmos: Three Thousand Years on the Shaman's Path* (New York: William Morrow and Co., 1993).

5. Carl Johan Calleman, *Solving the Greatest Mystery of Our Time: The Mayan Calendar* (London: Garev Publishing, 2001); Carl Johan Calleman, *The Mayan Calendar and the Transformation of Consciousness* (Rochester, Vt.: Bear & Co., 2004).

6. Michael D. Coe, *Breaking the Maya Code* (London: Thames and Hudson, 1992).

7. Linda Schele and David Freidel, *A Forest of Kings: The Untold Story of the Ancient Maya* (New York: Morrow, 1990), 57.

8. R. Wilk, "The Ancient Maya and the Political Present," *J. Anthropol. Res.* 41 (1985): 307–26.

9. New7Wonders Foundation, "The Official New 7 Wonders of the World," www .new7wonders.com/classic/en/n7w/results/ (accessed August 3, 2009).

10. See for instance the Ordinances of Tomas Lopez for the regimentation of the Indians in 1552 in Diego de Landa, *Yucatan Before and After the Conquest* (New York: Dover, 1978), 179.

11. Diego de Landa, *Yucatan Before and After the Conquest* (New York: Dover, 1978).

12. David Freidel, Linda Schele, and Joy Parker, *Maya Cosmos: Three Thousand Years on the Shaman's Path* (New York: William Morrow and Co., 1993).

13. Russell Thornton, "Decline of Population in Mexico," in *American Indian Holocaust and Survival: A Population History Since 1492–1991.* The Civilization of the American Indian Series (Norman: University of Oklahoma Press, 1987).

14. James Painter, "Task for New Guatemalan President: The New Centre-Left President of Guatemala, Alvaro Colom, Wants His Country to Be a Model of Social Democracy with a 'Mayan Face,'" BBC News, November 6, 2007, http:// news.bbc.co.uk/2/hi/americas/7081312.stm.

15. Bodil Liljefors Persson, "The Hitchhiker's Guide to the Maya Universe: An Exploration of the Books of Chilam Balam" (dissertation, Lund University, 1996); Ralph L. Roys, *The Book of Chilam Balam of Chumayel* (Washington, D.C.: Smithsonian Institution, 1933).

16. Tony Shearer, *Beneath the Moon and Under the Sun: A Poetic Re-Appraisal of the Sacred Calendar and the Prophecies of Ancient Mexico* (Santa Fe, N. Mex.: Sun Books, 1975).

17. Frank Waters, *Mexico Mystique: The Coming 6th World of Consciousness* (Chicago: Swallow Press, 1989).

18. Peter Balin, *The Flight of Feathered Serpent* (Wilmot, Wis.: Arcana, 1978).

19. Geoff Stray, personal communication.

20. Ralph Roys, *Book of Chilam Balam,* 42–63.

21. Carl Johan Calleman, *The Mayan Calendar and the Transformation of Consciousness* (Rochester, Vt.: Bear & Co., 2004).

22. For Carl Sagan's Cosmic Clock see Discovery Education, "The Universe in One Year," http://school.discoveryeducation.com/schooladventures/universe/itsawe-some/cosmiccalendar/page2.html (accessed August 3, 2009).

23. Carl Johan Calleman, *Solving the Greatest Mystery of Our Time: The Mayan Calendar* (London: Garev Publishing, 2001).

24. Lynn Margulis and Karlene Schwartz, *Five Kingdoms*, 3rd ed. (New York: W. H. Freeman and Co., 1998).

25. Frank E. Poirier, *Understanding Human Evolution* (Englewood Cliffs, N.J.: Prentice Hall, 1990), 103.

26. Philip Whitfield, *From So Simple a Beginning: The Book of Evolution* (New York: MacMillan, 1993) and Donald R. Prothero, *Evolution: What the Fossils Say and Why It Matters* (New York: Columbia University Press, 2007), 165. Whitfield gives 850 MYA and Prothero gives 800–900 MYA.

27. Helena Curtis and Sue Barnes, *Biology,* 5th ed. (New York: Worth, 1989), 1039. Footsteps were covered by ash from a volcanic eruption in Serengeti, which protected them for future study.

28. Frank E. Poirier, *Understanding Human Evolution* (Englewood Cliffs, N.J.: Prentice Hall, 1990), 103.

29. T. D. White, B. Asfaw, D. DeGusta et al., "Pleistocene *Homo sapiens* from Middle Awash, Ethiopia," *Nature* 423 (2003): 742–47, doi:10.1038/nature01669.

CHAPTER 3.
BASIC QUESTIONS REGARDING
BIOLOGICAL EVOLUTION

1. James Hutton, *Theory of the Earth; or an Investigation of the Laws Observable in the Composition, Dissolution, and Restoration of Land upon the Globe* (Transactions of the Royal Society of Edinburgh, 1785).

2. As pointed out by Eva Jablonka and Marion J. Lamb, Darwin talks about "Variability . . . from use and disuse" in *On the Origin of Species.* Eva Jablonka and Marion J. Lamb, *Evolution in Four Dimensions: Genetic, Epigenetic, Behavioral, and Symbolic Variation in the History of Life* (Cambridge, Mass.: MIT Press, 2006).

3. Eva Jablonka and Marion J. Lamb, *Evolution in Four Dimensions: Genetic, Epigenetic, Behavioral, and Symbolic Variation in the History of Life* (Cambridge, Mass.: MIT Press, 2006).

4. Charles Darwin, *The Origin of Species* (London: John Murray, 1859).

5. Richard Dawkins, *The God Delusion* (New York: Mariner Books, 2008).

6. National Academy of Sciences, Institute of Medicine, *Science, Evolution and Creationism,* www.nap.edu/catalog.php?record_id=11876 (accessed August 3, 2009).

7. National Research Council and National Academy of Sciences, *Science and Creationism: A View from the National Academy of Sciences,* 2nd ed. (Washington, D.C.: National Academy Press, 1999).

8. National Center for Science Education, Defending the Teaching of Evolution in the Public Schools, www.ncseweb.org/resources/news/2007/US/20_a_new_creationismevolution_po_4_4_2007.asp.

9. J. D. Watson and F. H. C. Crick, "A Structure for Deoxyribose Nucleic Acid," *Nature* 171 (1953): 737–38.

10. Stephen C. Meyer, "Evidence for Design in Physics and Biology: From the Origin of the Universe to the Origin of Life," in *Science and Evidence for Design in the Universe,* Michael J. Behe, William A. Dembski, and Stephen C. Meyer, eds. (San Francisco: Ignatius, 2000), 74–75.

11. E. Pennisi, "Working the (Gene Count) Numbers: Finally, a Firm Answer?" *Science* 316 (2007): 1113.

12. Donald R. Prothero, *Evolution: What the Fossils Say and Why It Matters* (New York: Columbia University Press, 2007), 78.

13. Robert A. Rohde and Richard A. Muller, "Cycles in Fossil Diversity," *Nature* 434 (2005): 208–10.

14. Michael Behe, *Darwin's Black Box: The Biochemical Challenge to Evolution* (New York: Free Press, 1996), 39–40.

15. Stephen Jay Gould, "The Pattern of Life's History," www.edge.org/3rd_culture/gould/gould_index.html (accessed August 3, 2009).

16. Stuart A. Kauffman, *The Origins of Order: Self-Organization and Selection in Evolution* (New York: Oxford University Press, 1993).

CHAPTER 4.
THE TREE OF LIFE HYPOTHESIS

1. M. Tegmark et al., "Cosmological Parameters from SDSS and WMAP," *Physical Review D* 69 (2004): 103501.

2. N. Dauphas, "The U/Th Production Ratio and the Age of the Milky Way from Meteorites and Galactic Halo Stars," *Nature* 435 (2005): 1203–05.

3. Michael D. Lemonick, "The Violence of Creation," *Time,* June 19, 1995, 48.

4. Peter D. Ward and Donald Brownlee, *Rare Earth: Why Complex Life Is Uncommon in the Universe* (New York: Copernicus Books, 2004), 47.

5. Ibid, 51.

6. Ibid, 97.

7. Lynn Margulis and Karlene V. Schwarz, *Five Kingdoms,* 3rd ed. (New York: W. H. Freeman, 1998).

8. Lesley J. Rogers and Richard J. Andrew, eds., *Comparative Vertebrate Lateralization* (Cambridge: Cambridge University Press, 2002).

9. Donald R. Prothero, *Evolution: What the Fossils Say and Why It Matters* (New York: Columbia University Press, 2007), 216–20.

10. Peter D. Ward and Donald Brownlee, *Rare Earth: Why Complex Life Is Uncommon in the Universe* (Katlenburg-Lindau, Germany: Copernicus Books, 2004), 47.

11. Carl Johan Calleman, *Solving the Greatest Mystery of Our Time: The Mayan Calendar* (London: Garev Publishing, 2001), 95–108.

12. Robert A. Rohde and Richard A. Muller, "Cycles in Fossil Diversity," *Nature* 434 (2005): 208–10.

13. J. John Sepkoski, "A Compendium of Fossil Marine Animal Genera," *Bulletins of American Paleontology* 363 (2002): 1–560.

14. B. S. Lieberman and A. L. Melott, "Considering the Case for Biodiversity Cycles: Re-Examining the Evidence for Periodicity in the Fossil Record," *PLoS ONE* 8 (April 2007), http://arxiv.org/pdf/0704.2896 (accessed August 3, 2009).

15. J. W. Valentine, A. G. Collins, and C. P. Meyer, "Morphological Complexity Increase in Metazoans," *Paleobiology* 20 (1994): 131–42.

16. Frank E. Poirier, *Understanding Human Evolution* (Englewood Cliffs, N.J.: Prentice Hall, 1990), 103.

17. M. Leakey, "Footprints Frozen in Time," *National Geographic* 155 (1979): 446–57.

18. Yves Coppens, *Le singe, l'Afrique et l'homme* (Paris: Librairie Arthème Fayard, 1983).

19. L. S. B. Leakey, P. V. Tobias, and J. R. Napier, "A New Species of Genus *Homo* from Olduvai Gorge," *Nature* 202 (1964): 7–9.

20. T. D. White, B. Asfaw, D. DeGusta et al., "Pleistocene *Homo sapiens* from Middle Awash, Ethiopia," *Nature* 423 (2003): 742–47.

21. "The Physical Characteristics of Humans," www.wsu.edu:8001/vwsu/gened/
learn-modules/top_longfor/phychar/culture-humans-2two.html (accessed
August 3, 2009).

22. Michael J. Benton, *When Life Nearly Died: The Greatest Mass Extinction of
All Time* (London and New York: Thames and Hudson, 2005).

23. K. J. Niklas, B. H. Tiffney, and A. H. Knoll, "Patterns in Vascular Land Plant
Diversification: An Analysis at the Species Level," in *Phanerozoic Profiles in
Macroevolution* (Princeton, N.J.: Princeton University Press, 1985), 97–128.

24. Nicholas Wade, ed., *The Science Times Book of the Brain* (New York: Lyons
Press, 1987), 150.

CHAPTER 5.
THE CONSTANTS OF NATURE ARE NOT FINE-TUNED
FOR LIFE BY ACCIDENT

1. Harold J. Morowitz, *Beginnings of Cellular Life: Metabolism Recapitulates
Biogenesis* (New Haven, Conn.: Yale University Press, 1992), 42.

2. C. H. Lineweaver, "An Estimate of the Age Distribution of Terrestrial Planets
in the Universe: Quantifying Metallicity as a Selection Effect," *Icarus* 151
(2001): 307–13.

3. Ibid.

4. D. R. Garnett and H. A. Kobulnicky, "Distance Dependence in the Solar
Neighborhood Age-Metallicity Relation," *Astrophysical Journal* 532 (2000):
1192–96.

5. C. H. Lineweaver, "An Estimate of the Age Distribution of Terrestrial Planets
in the Universe," *Icarus* 151 (2001): 307–13.

6. Matthew Hedman, *The Age of Everything: How Science Explores the Past*
(Chicago: University of Chicago Press, 2007), 142–62.

7. See for example Bruce Sterling, "Formerly Elegant Milky Way Galaxy
Downclassed As Cheap, Lousy 'Bar Galaxy,'" *Wired,* June 8, 2008, http://blog.
wired.com/sterling/2008/06/formerly-elegan.html (accessed August 3, 2009).

8. Matthew Hedman, *The Age of Everything: How Science Explores the Past*
(Chicago: University of Chicago Press, 2007), 210.

9. D. Clowe, M. Bradac, A. H. Gonzalez et al., "A Direct Empirical Proof of the
Existence of Dark Matter," *Astrophysical J. Lett.* 648 (2006): 109–13, e-print:
arXiv:astro-ph/0608407v1.

10. The word *tun* incidentally means "stone," since the Maya erected stone steles on tun shift celebrations. See Linda Schele and David Freidel, *A Forest of Kings: The Untold Story of the Ancient Maya* (New York: Morrow, 1990), 81.

11. See for instance "Bible Numbers," www.netrover.com/~numbers/lesson_2b .bible-codes.html (accessed August 3, 2009).

12. M. H. Hart, "The Evolution of the Atmosphere of the Earth," *Icarus* 33 (1978): 23–39.

13. A. Cazenave, "The Length of the Day," in *The Cambridge Atlas of Astronomy*, Jean Audouze and Guy Israel, eds., 3rd ed. (Cambridge: Cambridge University Press, 1994), 59.

14. See for instance David Lindorff, *Pauli and Jung: The Meeting of Two Great Minds* (Wheaton, Ill.: Quest Books, 2004) or C. A. Meier, ed., *Atom and Archetype: The Pauli/Jung Letters, 1932–1958* (Princeton, N.J.: Princeton University Press, 1992).

15. Richard P. Feynman, *QED: The Strange Theory of Light and Matter* (Princeton, N.J.: Princeton University Press, 1985), 129.

16. R. Heyrovska and S. Narayan, "Fine-structure Constant, Anomalous Magnetic Moment, Relativity Factor and the Golden Ratio that Divides the Bohr Radius," doi:arXiv:physics/0509207; see http://arxiv.org/abs/physics/0509207.

17. See www.goldennumber.net.

18. Mario Livio, *The Golden Ratio: The Story of PHI, the World's Most Astonishing Number* (New York: Broadway, 2003).

19. R. Heyrovska, "The Golden Ratio, Ionic and Atomic Radii and Bond Lengths," *Mol. Phys.* 103 (2005): 877–82; and R. Heyrovska, "Golden Sections of Inter-Atomic Distances as Exact Ionic Radii of Atoms," *Naturepreceedings*, doi:hdl:10101/npre.2009.2929.1.

CHAPTER 6.
THE WIDER CONTEXT OF BIOLOGICAL EVOLUTION

1. Life in the Fast Lane, "World's Oldest Ritual Discovered: Serpent Worship," www.lifeinthefastlane.ca/worlds-oldest-ritual-discovered-serpent-worship/ weird-science (accessed August 3, 2009).

2. M. S. Kim and J. Cho, "Teleporting the Quantum State to Distant Matter," *Science* 323 (2009): 469–70.

3. Rupert Sheldrake, *Dogs That Know When Their Owners Are Coming Home:*

And Other Unexplained Powers of Animals (New York: Three Rivers Press, 2000).

4. James Binney and Scott Tremaine, *Galactic Dynamics,* Princeton Series in Astrophysics (Princeton, N.J.: Princeton University Press, 1987), 17.

5. Mark Morris, "The Milky Way," *The World Book Encyclopedia,* 2002, vol. 13: 551.

6. Jesper Sommer-Larsen, Copenhagen Observatory, personal communication.

7. Richard J. Andrew and Lesley J. Rogers, "The Nature of Lateralization in Tetrapods" in *Comparative Vertebrate Lateralization,* ed. Lesley J. Rogers and Richard J. Andrew (Cambridge: Cambridge University Press, 2002), 94–125.

8. David Suzuki and Peter Knudtson, *Wisdom of the Elders: Native and Scientific Ways of Knowing about Nature* (Toronto: Douglas and McIntyre, 2006), 23.

9. Peter D. Ward and Donald Brownlee, "The Surprising Importance of Plate Tectonics," in *Rare Earth: Why Complex Life Is Uncommon in the Universe* (New York: Copernicus Books, 2004), 191–220.

10. B. Haq, J. Hardenbol, and P. Vail, "Chronology of Fluctuating Sea Levels Since the Triassic," *Science* 235 (1987): 1156–67.

11. I. W. D. Dalziel, "Earth before Pangaea," *Scientific American* 272 (1995): 58–63.

12. T. H. Torsvik, "The Rodinia Jigsaw Puzzle," *Science* 300 (2003): 1379–81.

13. I. W. D. Dalziel, "Earth before Pangaea," *Scientific American* 272 (1995): 58–63.

14. Guillermo Gonzalez and Jay Richards, *The Privileged Planet: How Our Place in the Cosmos Is Designed for Discovery* (Washington, D.C.: Regnery Publishing, 2004), plate 6.

15. A. Jephcoat and K. Refson, "Earth Science: Core Beliefs," *Nature* 413 (2001): 27–30, doi:10.1038/35092650.

16. E. J. Garnero and A. K. McNamara, "Structure and Dynamics of Earth's Lower Mantle," *Science* 320 (2008): 626–28.

17. Michael J. Benton, *When Life Nearly Died: The Greatest Mass Extinction of All Time* (London and New York: Thames and Hudson, 2005).

18. Peter D. Ward and Donald Brownlee, "The Surprising Importance of Plate Tectonics," in *Rare Earth: Why Complex Life Is Uncommon in the Universe* (Katlenburg-Lindau, Germany: Copernicus Books, 2004), 191–220.

19. M. J. Benton, "The Red Queen and the Court Jester: Species Diversity and the Role of Biotic and Abiotic Factors through Time," *Science* 323 (2009): 728–32.

20. Donald R. Prothero, *Evolution: What the Fossils Say and Why It Matters* (New York: Columbia University Press, 2007).

21. Michael Sanderson of the University of Arizona, quoted in Carl Zimmer, "Crunching the Data for the Tree of Life," *New York Times,* February 10, 2009.

22. The Extrasolar Planets Encyclopaedia, http://exoplanet.eu/.

23. D. S. McKay, E. K. Gibson Jr., K. L. Thomas-Keprta et al., "Search for Past Life on Mars: Possible Relic Biogenic Activity in Martian Meteorite ALH84001," *Science* 273 (1996): 924–30, doi:10.1126/science.273.5277.924.

24. T. Satyanarayana, C. Raghukumar, and S Shivaji, "Extremophilic Microbes: Diversity and Perspectives," *Current Science* 89 (2005): 78–90.

25. Guillermo Gonzalez and Jay Richards, *The Privileged Planet: How Our Place in the Cosmos Is Designed for Discovery* (Washington, D.C.: Regnery Publishing, 2004), app. A: 337–42.

CHAPTER 7.
THE BIOCHEMICAL BASIS OF BIOLOGICAL EVOLUTION

1. S. L. Miller, "Production of Amino Acids Under Possible Primitive Earth Conditions," *Science* 117 (1953): 528–29, doi:10.1126/science.117.3046.528.

2. Noam Lahav, *Biogenesis: Theories of Life's Origin* (New York: Oxford University Press, 1999), 240.

3. M. E. Beleza Yamagishi and A. I. Shimabukuro, "Nucleotide Frequencies in Human Genome and Fibonacci Numbers," *Bull. Math. Biol.* 70 (2008): 643–53.

4. J. Cairns, J. Overbaugh, and S. Miller, "The Origin of Mutants," *Nature* 335 (1988): 142–45.

5. R. S. Galhardo, P. J. Hastings, and S. M. Rosenberg, "Mutation as a Stress Response and the Regulation of Evolvability," *Critical Reviews in Biochemistry and Molecular Biology* 42 (2007): 399–435, doi:10.1080/10409230701648502.

6. S. Arrhenius, "Panspermy: The Transmission of Life from Star to Star," *Scientific American* 196 (1907): 196.

7. F. H. C. Crick and L. E. Orgel, "Directed Panspermia," *Icarus* 19 (1973): 341–46.

8. David R. Williams, NASA, "Evidence of Ancient Martian Life in Meteorite ALH84001?" http://nssdc.gsfc.nasa.gov/planetary/marslife.html (accessed August 3, 2009).

9. Guardians of the Millennium, "Extremophiles," www.theguardians.com/ Microbiology/gm_mbm04.htm (accessed August 3, 2009).

10. N. H. Sleep, K. J. Zahnle, J. F. Kasting, and H. J. Morowitz, "Annihilation of Ecosystems by Large Asteroid Impacts on the Early Earth," *Nature* 342 (1989): 139–42, doi:10.1038/342139a0.

11. Peter D. Ward and Donald Brownlee, *Rare Earth: Why Complex Life Is Uncommon in the Universe* (New York: Copernicus Books, 2004), 61.

12. N. H. Sleep, K. J. Zahnle, J. F. Kasting, and H. J. Morowitz, "Annihilation of Ecosystems by Large Asteroid Impacts on the Early Earth," *Nature* 342 (1989): 139–42, doi:10.1038/342139a0.

13. D. S. McKay, E. K. Gibson, K. L. Thomas-Keprta et al., "Search for Past Life on Mars: Possible Relic Biogenic Activity in Martian Meteorite ALH84001," *Science* 273 (1996): 924–30, doi:10.1126/science.273.5277.924.

14. K. Nakamura-Messenger, "Organic Globules in the Tagish Lake Meteorite: Remnants of the Protosolar Disc," *Science* 314 (2006): 1439–42.

15. Astrobiology Web, "Microbe from Depths Takes Life to Hottest Known Limit," www.astrobiology.com/news/viewpr.html?pid=12337 (accessed August 3, 2009).

16. C. Schleper, G. Pühler, B. Kühlmorgen, and W. Zillig, "Life at Extremely Low pH," *Nature* 375 (1995): 741–42.

17. G. Wächtershäuser, "Evolution of the First Metabolic Cycles," *Proc Natl Acad Sci USA* 87 (1990): 200.

CHAPTER 8.
A NEW THEORY OF BIOLOGICAL EVOLUTION

1. C. A. Thomas, "The Genetic Organization of Chromosomes," *Annual Review of Genetics* 5 (1971): 237–56.

2. John W. Kimball, "Genome Sizes," http://users.rcn.com/jkimball.ma.ultranet/ BiologyPages/G/GenomeSizes.html (accessed August 3, 2009).

3. E. Pennisi, "Working the (Gene Count) Numbers: Finally, a Firm Answer?" *Science* 316 (2007): 1113, doi:10.1126/science.316.5828.1113a.

4. D. E. Wildman, M. Uddin, G. Liu, L. I. Grossman, and M. Goodman, "Implications of Natural Selection in Shaping 99.4% Nonsynonymous DNA Identity between Humans and Chimpanzees: Enlarging Genus *Homo*," *Proc. Natl. Acad. Sci. USA* 100 (2003): 7181–88.

5. A. Varki and T. K. Altheide, "Comparing the Human and Chimpanzee Genomes: Searching for Needles in a Haystack," *Genome Res.* 15 (2005): 1746–58.

6. E. B. Lewis, "A Gene Complex Controlling Segmentation in *Drosophila*," *Nature* 276 (1978): 565–70.

7. S. Hameroff, *Ultimate Computing: Biomolecular Consciousness and Nanotechnology* (Holland: Elsevier Science, 1987).

8. T. Boveri, "Zellenstudien II. Die Befruchtung und Teilung des Eies von Ascaris megalocephala," *Jena Zeit. Naturw.* 22 (1888): 685–882.

9. See the phylogenetic tree in J. Beisson and M. Wright, "Basal Body/Centriole Assembly and Continuity," *Current Opinion in Cell Biology* 15 (2003): 96–104.

10. Dennis N. Wheatley, *The Centriole* (Holland: Elsevier, 1982).

11. See for instance the discussion in A. M. Preble, T. M. Giddings Jr., and S. K. Dutcher, "Basal Bodies and Centrioles: Their Function and Structure" in *The Centrosome in Cell Replication and Early Development, Current Topics in Developmental Biology,* vol. 49, Robert E. Palazzo and Gerald P. Schatten, eds. (San Diego, Calif.: Academic Press, 2000).

12. J. L. Feldman, S. Geimer, and W. F. Marshall, "The Mother Centriole Plays an Instructive Role in Defining Cell Geometry," *PLoS Biol* 5 (6) (2007): e149, doi:10.1371/journal.pbio.0050149.

13. C. R. Cowan and A. A. Hyman, "Centrosomes Direct Cell Polarity Independently of Microtubule Assembly in *C. elegans* Embryos," *Nature* 431 (2004): 92–96, doi:10.1038/nature02825.

14. S. Nonaka, S. Yoshiba, D. Watanabe et al., "De Novo Formation of Left-Right Asymmetry by Posterior Tilt of Nodal Cilia," *PLoS Biol* 3 (2005): 1467–72.

15. J. Y. Li and C. F. Wu, "New Symbiotic Hypothesis on the Origin of Eukaryotic Flagella," *Naturwissenschaften* 92 (2005): 305–9.

16. J. L. Feldman, S. Geimer, and W. F. Marshall, "The Mother Centriole Plays an Instructive Role in Defining Cell Geometry," *PLoS Biol* 5 (6) (2007): e149, doi:10.1371/journal.pbio.0050149.

17. A. I. Gotlieb, L. McBurnie-May, L. Subrahmanyan, and V. I. Kalnins, "Distribution of Microtubule Organizing Centers in Migrating Sheets of Endothelial Cells," *The Journal of Cell Biology* 91 (1981): 589–94.

18. For a good overview of cilia and flagella, see Gerald Karp, *Cell and Molecular Biology,* 2nd ed. (New York: John Wiley and Sons, 1999), 366–74.

19. R. Basto, J. Lau, T. Vinogradova, A. Gardiol et al., "Flies without Centrioles," *Cell* 125 (2006): 1375–86.

20. E. Karsenti and I. Vernos, "The Mitotic Spindle: A Self-Made Machine," *Science* 294 (2001): 543.

21. T. Küntziger and M. Bornens, "The Centrosome and Parthenogenesis," in *The Centrosome in Cell Replication and Early Development, Current Topics in Developmental Biology,* vol. 49, ed. Robert E. Palazzo and Gerald P. Schatten (San Diego, Calif.: Academic Press, 2000), 1–25.

CHAPTER 9.
A BIOLOGY WITH SOUL

1. G. Albrecht-Buehler, "Rudimentary Form of Cellular 'Vision,'" *Proc. Natl. Acad. Sci. USA* 89 (1992): 8288–92.

2. G. Albrecht-Buehler, "Functions and Formation of Centrioles and Basal Bodies," in *The Centrosome,* V. I. Kalnins, ed. (Academic Press, 1993), 69–102.

3. Ralph L. Roys, "The Creation of the World," in *The Book of Chilam Balam of Chumayel,* chap. 10 (Norman: University of Oklahoma Press, 1967).

4. A. Cho, "A Singular Conundrum: How Odd Is Our Universe?" *Science* 317 (2007): 1848–50.

5. J. C. Lassiter, "Constraints on the Coupled Thermal Evolution of the Earth's Core and Mantle, the Age of the Inner Core, and the Origin of the $^{186}Os/^{188}Os$ 'Core Signal' in Plume-derived Lavas," *Earth and Planetary Science Letters* 250 (2006): 306–17.

6. A. B. Belonoshko, N. V. Skorodumova, A. Rosengren, and B. Johansson, "Elastic Anisotropy of Earth's Inner Core," *Science* 319 (2008): 797–800.

7. S. Hameroff, *Ultimate Computing: Biomolecular Consciousness and Nanotechnology* (Holland: Elsevier Science, 1987).

8. S. Hameroff, A. Nip, M. Porter et al., "Conduction Pathways in Microtubules, Biological Quantum Computation, and Consciousness," *Biosystems* 64 (2002): 149–68.

9. Drunvalo Melchizedek, *The Ancient Secret of the Flower of Life,* vol. 2 (Flagstaff, Ariz.: Light Technology Publications, 2000).

10. Priya Hemenway, *Divine Proportion: Phi in Art, Nature, and Science* (New York: Sterling, 2005).

11. Johann Wolfgang von Goethe, quoted in M. J. Denton, C. J. Marshall, and

M. Legge, "The Protein Folds as Platonic Forms: New Support for Pre-Darwinian Evolution by Natural Law," *J. Theor Biol.* 219 (2002): 325–42. This has an interesting discussion of Platonic forms in biology.

12. See for example Wynne Brown, Discovery Health, "Alternative Medicine Goes Mainstream," http://health.discovery.com/centers/althealth/medtrends/medtrends.html (accessed August 3, 2009).

13. R. Mays and S. Mays, "Phantom Limb 'Touch' Suggests That a 'Mind-Link' Extends beyond the Physical Body and Can Interact with Reality, Producing Physiological Sensations," Abstract 158, *Toward a Science of Consciousness* (April 8–12, 2008).

14. G. Albrecht-Buehler, "Changes of Cell Behavior by Near-Infrared Signals," *Cell Motil Cytoskeleton* 32 (1995): 299–304.

15. A. I. Gotlieb, L. McBurnie-May, L. Subrahmanyan, and V. I. Kalnins, "Distribution of Microtubule Organizing Centers in Migrating Sheets of Endothelial Cells," *The Journal of Cell Biology* 91 (1981): 589–94.

16. N. Korzeniewski, L. Zheng, R. Cuevas, J. Parry, P. Chatterjee, B. Anderton, A. Duensing, K. Münger, and S. Duensing, "Cullin 1 Functions as a Centrosomal Suppressor of Centriole Multiplication by Regulating Polo-like Kinase 4 Protein Levels," *Cancer Resarch* 69 (2009): 6668, doi:10.1158/0008-5472.CAN-09-1284.

17. K. Fukasawa, "Oncogenes and Tumour Suppressors Take On Centrosomes," *Nature Rev. Cancer* 7 (2007): 911–24.

18. S. R. Hameroff. "A New Theory of the Origin of Cancer: Quantum Coherent Entanglement, Centrioles, Mitosis, and Differentiation," *Biosystems* 77 (2004): 119–36.

19. B. N. Ames, W. E. Durston, E. Yamasaki, and F. D. Lee, "Carcinogens Are Mutagens: A Simple Test System Combining Liver Homogenates for Activation and Bacteria for Detection," *Proc. Natl. Acad. Sci. USA* 70 (1973): 2281–85.

20. C. J. Calleman et al., "Monitoring and Risk Assessment by Means of Alkyl-Groups in Hemoglobin in Persons Occupationally Exposed to Ethylene Oxide," *J. Environ. Pathol. Toxicol.* 2 (1978): 427–42.

21. C. J. Calleman, "The Metabolism and Pharmacokinetics of Acrylamide: Implications for Mechanisms of Toxicity and Human Risk Estimation," *Drug Metabol. Rev.* 28 (1996): 527–90.

22. L. Susskind, *The Cosmic Landscape: String Theory and the Illusion of Intelligent Design* (New York: Little, Brown, 2006), 17.

23. See for example Discovery Institute, www.intelligentdesign.org.

24. Stephen Hawking, *A Brief History of Time* (New York: Bantam, 1988).

25. G. Schatten, "The Centrosome and Its Mode of Inheritance: The Reduction of the Centrosome during Gametogenesis and Its Restoration during Fertilization," *Dev. Biol.* 165 (1994): 299–335.

26. T. Boveri, "Zellenstudien II. Die Befruchtung und Teilung des Eies von Ascaris megalocephala," *Jena Zeit. Naturw.* 22 (1888): 685–882.

Bibliography

Audouze, Jean, and Guy Israel, eds. *The Cambridge Atlas of Astronomy,* 3rd ed. Cambridge: Cambridge University Press, 1994.

Barrow, John, and Frank Tipler. *The Anthropic Cosmological Principle.* Gloucestershire, UK: Clarendon Press, 1986.

Behe, Michael J. *Darwin's Black Box: The Biochemical Challenge to Evolution.* New York: Free Press, 2006.

———. *The Edge of Evolution: The Search for the Limits of Darwinism.* New York: Free Press, 2007.

———. *Science and Evidence for Design in the Universe.* Edited by William A. Dembski and Stephen C. Meyer. San Francisco: Ignatius Press, 2000.

Bentov, Itzhak. *Stalking the Wild Pendulum: On the Mechanics of Consciousness.* Rochester, Vt.: Destiny Books, 1988.

Berg, Lasse. *Gryning över Kalahari.* Stockholm: Ordfront Förlag, 2005.

Bianki, V. L., and E. B. Fillipova. *Sex Differences in Lateralization in the Animal Brain.* Amsterdam: Harwood Academic, 2000.

Bohm, David. *Wholeness and the Implicate Order.* London: Routledge, 2002.

Brockman, John, ed. *Intelligent Thought: Science Versus the Intelligent Design Movement.* New York: Vintage Books, 2006.

Calleman, Carl Johan. *The Mayan Calendar and the Transformation of Consciousness.* Rochester, Vt.: Bear & Co., 2004.

———. *Solving the Greatest Mystery of Our Time: The Mayan Calendar.* London: Garev Publishing, 2001.

Cambridge Encyclopedia of Human Evolution. Cambridge: Cambridge University Press, 1992.

Campbell, Neil A., Jane B. Reece, and Lawrence G. Mitchell. *Biology.* 5th ed. Menlo Park, Calif.: Addison-Wesley, 1999.

Coe, Michael D. *Breaking the Maya Code.* London: Thames and Hudson, 1992.

———. *The Maya.* New York: Thames and Hudson, 1993.

Crick, Francis. *Astonishing Hypothesis: The Scientific Search for the Soul.* New York: Scribner, 1995.

Curtis, Helena, and Sue Barnes. *Biology.* 5th ed. New York: Worth, 1989.

Darwin, Charles. *On the Origin of Species by Means of Natural Selection, or the Preservation of Favoured Races in the Struggle for Life.* London: Collier-MacMillan, 1962 [1859].

Dawkins, Richard. *The Blind Watchmaker.* New York: W. W. Norton, 1986.

———. *The Selfish Gene.* London: Paladin Books, 1978.

Davies, Paul. *The Mind of God: The Scientific Basis for a Rational World.* London: Orion Production, 1992.

de Duve, Christian. *Vital Dust: The Origin and Evolution of Life on Earth.* New York: Basic Books, 1995.

Denton, Michael. *Evolution: A Theory in Crisis.* London: Burnett Books, 1985.

Dowd, Michael. *Thank God for Evolution: How the Marriage of Science and Religion Will Transform Your Life and Our World.* New York: Viking Adult, 2008.

Edwards, Lawrence. *The Vortex of Life: Nature's Patterns in Space and Time.* Edinburgh: Floris Books, 1993.

Ferguson, Kitty. *The Fire in the Equations: Science, Religion, and the Search for God.* West Conshohocken, Pa.: Templeton Foundation Press, 2004.

Freidel, David, Linda Schele, and Joy Parker. *Maya Cosmos.* New York: Harper Paperbacks, 1995.

Gardner, James N. *Biocosm: The New Scientific Theory of Evolution: Intelligent Life Is the Architect of the Universe.* Makawao, Hawaii: Inner Ocean, 2003.

Gonzalez, Guillermo, and Jay Richards. *The Privileged Planet: How Our Place in the Cosmos Is Designed for Discovery.* Washington, D.C.: Regnery, 2004.

Hameroff, Stuart. *Ultimate Computing: Biomolecular Consciousness and Nanotechnology.* Holland: Elsevier Science, 1987.

Hawking, Stephen W. *A Brief History of Time: From the Big Bang to Black Holes.* London: Bantam Press, 1988.

Hedman, Matthew. *The Age of Everything: How Science Explores the Past.* Chicago: University of Chicago Press, 2008.

Hemenway, Priya. *Divine Proportion: Phi in Art, Nature, and Science.* New York: Sterling, 2005.

The Holy Bible, New International Version. London: Hodder and Stoughton, 1978.

Jablonka, Eva, and Marion J. Lamb. *Evolution in Four Dimensions: Genetic, Epigenetic, Behavioral, and Symbolic Variation in the History of Life.* Cambridge, Mass.: MIT Press, 2006.

Karp, Gerald. *Cell and Molecular Biology.* 2nd ed. New York: John Wiley and Sons, 1999.

Kauffman, Stuart A. *The Origins of Order: Self-Organization and Selection in Evolution.* New York: Oxford University Press, 1993.

Koestler, Arthur *The Act of Creation.* London: Penguin, 1990.

———. *The Case of the Midwife Toad.* England: Hutchinson, 1971.

Lahav, Noam. *Biogenesis: Theories of Life's Origin.* New York: Oxford University Press, 1999.

Lipton, Bruce H. *The Biology of Belief: Unleashing the Power of Consciousness, Matter, and Miracles.* Carlsbad, Calif.: Hay House, 2008.

Livio, Mario. *The Golden Ratio: The Story of PHI, the World's Most Astonishing Number.* New York: Broadway, 2003.

Lovelock, James. *The Ages of Gaia: A Biography of Our Living Earth.* New York: Oxford University Press, 1990.

McFadden, Johnjoe. *Quantum Evolution: How Physics' Weirdest Theory Explains Life's Biggest Mystery.* New York: W. W. Norton and Company, 2002.

McKee, Jeffrey K., Frank E. Poirier, and Scott W. McGraw. *Understanding Human Evolution.* Englewood Cliffs, N.J.: Prentice Hall, 2004.

McTaggart, Lynne. *The Field: The Quest for the Secret Force of the Universe.* Updated ed. New York: Harper Paperbacks, 2008.

Milton, Richard. *Shattering the Myths of Darwinism.* Rochester, Vt.: Park Street Press, 2000.

Morowitz, Harold J. *Beginnings of Cellular Life: Metabolism Recapitulates Biogenesis.* New Haven, Conn.: Yale University Press, 1992.

———. *The Emergence of Everything: How the World Became Complex.* New York: Oxford University Press, 2004.

Nigg, Erich A. *Centrosomes in Development and Disease.* Germany: Wiley-VCH, 2004.

Olsen, Scott. *The Golden Section: Nature's Greatest Secret.* New York: Walker, 2006.

Palazzo, Robert E., and Gerald P. Schatten, eds. *The Centrosome in Cell Replication and Early Development.* Current Topics in Developmental Biology, vol. 49. San Diego, Calif.: Academic Press, 2000.

Parker, Steve. *The Dawn of Man.* Edited by Michael Day. New York: Crescent Books, 1992.

Peat, David. *Synchronicity: The Bridge Between Matter and Mind.* New York: Bantam, 1987.

Penrose, Roger. *Shadows of the Mind: A Search for the Missing Science of Consciousness.* New York: Oxford University Press, 1996.

Perez, Jean-Claude. *Codex biogenesis: 13 codes et harmonies de l'ADN.* Tome 1, Du Génome vers l'Atome. Edited by Marco Pietteur. Liege, France: Résurgence, 2009.

Perez, Jean-Claude. *L'ADN décrypté: La découverte et les preuves du langage caché de l'ADN.* Edited by Marco Pietteur. Liege, France: Résurgence, 1997.

Poirier, Frank E. *Understanding Human Evolution.* Englewood Cliffs, N.J.: Prentice Hall, 1990.

Prigogine, Ilya, and Isabelle Stengers. *Order out of Chaos.* New York: Bantam Books, 1984.

Prothero, Donald R. *Evolution: What the Fossils Say and Why It Matters.* New York: Columbia University Press, 2007.

Rees, Martin. *Just Six Numbers: The Deep Forces That Shape the Universe.* New York: Basic Books, 2001.

Rice, Prudence M. *Maya Political Science: Time, Astronomy, and the Cosmos.* Austin: University of Texas Press, 2004.

Rogers, Lesley J., and Andrew Richard, eds. *Comparative Vertebrate Lateralization.* Cambridge: Cambridge University Press, 2002.

Roys, Ralph L. *The Book of Chilam Balam of Chumayel.* Norman: University of Oklahoma Press, 1967.

Ruse, Michael. *Darwinism Defended.* N.J.: Addison-Wesley, 1982.

Schele, Linda, and David Freidel. *A Forest of Kings: The Untold Story of the Ancient Maya.* New York: Morrow, 1990.

Sheldrake, Rupert. *A New Science of Life: The Hypothesis of Formative Causation.* London: Paladin, 1987.

Talbot, Michael. *The Holographic Universe.* New York: Harper Perennial, 1992.

Tedlock, Dennis, trans. *Popol Vuh: The Mayan Book of the Dawn of Life*. New York: Simon and Schuster, 1985.

Umana, John. *Creation: Towards a Theory of All Things*. Booksurge, 2005.

Ward, Peter, and Donald Brownlee. *Rare Earth: Why Complex Life Is Uncommon in the Universe*. New York: Copernicus Books, 2004.

Watson, James D., Tania A. Baker, Stephen P. Bell, Alexander Gann, Michael Levine, and Richard Losick. *Molecular Biology of the Gene,* 5th ed. Upper Saddle River, N.J.: Benjamin Cummings, 2003.

Weinberg, Steven. *The First Three Minutes*. New York: Basic Books, 1993.

Wheatley, Dennis. N. *The Centriole*. Holland: Elsevier, 1982.

Whitfield, Philip. *From So Simple a Beginning: The Book of Evolution*. New York: MacMillan, 1993.

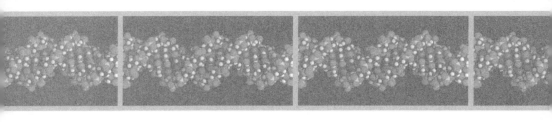

About the Author

Carl Johan Calleman holds a Ph.D. in Physical Biology from the University of Stockholm and was a senior researcher of environmental health at the University of Washington in Seattle from 1986 to 1993. He has served as a cancer expert for the World Health Organization and lectured at scientific institutes around the world, ranging from MIT to the Chinese Academy of Preventive Medicine in Beijing. As a toxicologist and chemist, his research articles have been quoted some 1,500 times in the scientific literature.

In addition to his scientific work, Calleman is an internationally recognized authority in the studies of the Mayan calendar. He began his studies on the Mayan calendar in 1979 and is the author of *Solving the Greatest Mystery of Our Time: The Mayan Calendar* and *The Mayan Calendar and the Transformation of Consciousness*. He lectures throughout the world and has made numerous appearances in television, radio, and the Internet.

Author's website
www.calleman.com

Web Resources about the Mayan calendar
www.maya-portal.net

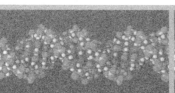

Index